INDUSTRIAL DESIGN – NEW FRONTIERS

Edited by **Denis A. Coelho**

Industrial Design – New Frontiers
Edited by Denis A. Coelho

Published by InTech
Janeza Trdine 9, 51000 Rijeka, Croatia

Publishing Process Manager Alenka Urbancic

Technical Editor Teodora Smiljanic

Cover Designer InTech Design Team

Additional hard copies can be obtained from orders@intechopen.com

Industrial Design – New Frontiers, Edited by Denis A. Coelho
p. cm.
ISBN 978-953-307-622-5

For EU product safety concerns:
IN TECH d.o.o., Prolaz Marije Krucifikse Kozulić 3, 51000 Rijeka, Croatia,
info@intechopen.com or visit our website at intechopen.com.

Meet the editor

Professor Denis A. Coelho's main teaching contribution has been stimulating the formation and initial operation of graduate courses in Industrial Design and MA in Industrial Design Engineering at UBI. He develops scientific activity within product design methodologies and has also conducted research in the field of ergonomic design and human requirements on technology, developed extensively in his doctoral work. Professor Coelho collaborates with several international research institutions in the areas of product design and ergonomics. The key area of his scientific activity is focused on the Human Interface located between technology and humans, which is characterized by an interdisciplinary nature, with technical and technological bias favouring an ergonomic perspective. His scientific activity also extends to product design methodologies, normalization applied to industrial management systems and production strategy in international networks. He is editor in chief of the International Journal of Human Factors and Ergonomics.

Contents

Preface

A new breed of modern designers is emerging. These non-traditional industrial designers work across disciplines, understand human beings, business and technology thus bridging the gap between customer needs and technological advancement of tomorrow. Whether by using unique manufacturing processes like rapid prototyping and laser cutting or by celebrating beautiful materials like organics, concrete or silicon, they are successfully evolving industrial design according to user's needs and engagement with their product. This book uncovers prospective designer techniques and methods of a new age of industrial design, in which practitioners strive to construct simple and yet complex products of the future.

The nine chapters in this book report on the novel frontiers of the new era of industrial design and how these are being tackled, spanning over the design process, illustration of the use of new technologies in design and the advancement of culturally inspired design.

The first section of the book is on design and new technologies, covering the themes of technology as a determinant of object shape, knowledge-based engineering for supporting design and product instructions in the digital age. The second section deals with culturally inspired design from both a practice and an educational viewpoint. The last section covers several approaches to innovation in the design process, including biologically inspired design, product design with embodiment design viewed in a new perspective, disruptive innovative designs, and an example of the application of TRIZ principles to design.

The diverse perspectives taken by the authors of this book ensure stimulating reading and will assist readers in leaping forward in their own practice of industrial design and in preparing new design research that is relevant and aligned with the current challenges of this fascinating field.

Professor Denis A. Coelho
Human Technology Group (Director)
Masters Program in Industrial Design Engineering (Coordinator)
Product Development and Prototyping Laboratory (Head)
Dept. Electromechanical Engineering, University of Beira Interior,
Covilhã, Portugal

Professor Abir Mullick
School of Industrial Design
College of Architecture
Georgia Institute of Technology
Atlanta, Georgia
United States of America

Part 1

Design and New Technologies

1

Technology as a Determinant of Object Shape

Denis A. Coelho and Filipe A. A. Corda
Universidade da Beira Interior
Portugal

1. Introduction

In a globalized world, we are confronted everywhere with the encouragement to consumption, the purchase of goods and services. This is something characteristic of the consumer society in which we live, whether it is caused by consumption needs and their satisfaction or because saturation has already set in, or it may even be caused by social statement (Baudrillard 1995).

The intention of this chapter is to focus on the study of technology as a determinant of the shape of products that incorporate technology. The acceptance of a product by consumers, and how important the shape and look of it is to the success of the product, has been previously studied (Bloch 1995). Research has also been carried out to study the determinants of the shape of products (Crilly, Moultrie & Clarkson 2009), which shows that several aspects determine the final shape of a product. The form development process is driven by the designers' efforts to guide or constrain the way in which the product will be experienced, and the success of the final design may be determined by the degree of correspondence between designer intent and consumer response (Crilly, Moultrie & Clarkson 2009). This chapter does not intend to focus on all determinants of product shape and does not aspire to focus on the acceptance of a product on the market or to reflect on the role of the personal taste of the industrial designer in determining the shape of the product. The aim of this chapter is to study the influence of one aspect in particular, technology as a determinant of the form of products, for products that embed technology.

The study of materials and their development also fits this line of inquiry, since there is a synergy between the development of technology and of materials. This historical relationship, while not the main focus of this work, is also of interest and merits some comments. The advent of new materials and the development and improvement of others makes it possible to improve the application of new technologies and the consequent development of new products. Man, machine, materials and production are closely linked in modern industry and this link is becoming increasingly strong. Advanced materials are critical to many new technological applications, since the latter depend strongly on the advances of the former (i.e. the high-speed train, Maglev, is based on technology already developed and tested but its large-scale implementation awaits improvements in materials technology so that cryogenic preservation can be maintained economically, so that it may be possible to create the magnetic levitation and consequent propulsion of the vehicle).

Mastering the state of the art in materials and technological solutions offers vast opportunities, due largely to a greater understanding and greater control of their basic characteristics. As such, new materials play an important role in the development of innovative technologies. For example, without knowledge of materials such as quartz crystal or piezoelectric ceramics, the production of energy that occurs with the deformation of these materials could not be put to use. Knowledge about materials catalyzes more technological knowledge. A recently created metallic material, with platinum-based nano-pores, expands and contracts under the action of an electric current thereby converting electrical energy into mechanical energy and vice versa, depending on the state of the material. In conclusion, materials in general and bio-mimetic nano-materials in particular, form together with intelligent materials and organic polymers, among other materials, a wide range of examples of materials whose importance is recognized in the field of technology and product design.

The approach that is central to this chapter is bounded by the larger process of industrial design, where it gives a contribution that may take place at the stage of concept generation as well as the stage of detailed design. The bounding process of design that is considered is in line with the report of Lewis and Bonollo (2002). These authors performed an experimental investigation to unveil the design skills most influential to professional success, in order to have design education adequately train students in those skills. In order to structure their research, these authors harnessed a five stage operational process of design, based on selected literature of their choice (Hales 1991). This process is comprised of five sub-ordinate processes (Table 1).

Product development is part of any company's industrial innovation process (Roozenburg & Eekels 1995). Industrial innovation includes all activities preceding the launch of a new product into the marketplace, such as basic and applied research, design and development, market research, production, distribution and sales. The development of new technological possibilities has triggered the search for applications, which in many cases has led to the unveiling of new products. In such cases it is not uncommon that the design process instead of centring on the user and the potential market is driven by the search for technological applications.

In practice, technology influenced design may lead the way, in some cases, to a simplification of the design process depicted in Table 1, as little room is given to new concept generation, since the concept is determined by the application of technology to enable a particular functionality and as such, circumvents the search for new ideas, and promotes the continuation of a particular product archetype. Such archetype may well be tied to market requirements and perceived consumer acceptance, as well as to technological constraints and salient features that hamper shape alterations. What is proposed in this chapter is to study the influence of technology as a determinant of the final shape of a product that incorporates technology. This is intended to demonstrate how this is a key aspect for the possible deconstruction of product archetypes which have endured for many years. To this end, some new product concepts are presented that incorporate emerging technologies, reflecting a change of form, distinct from the existing hitherto.

This chapter seeks to demonstrate the influence of technology in the form of the product. It also seeks to unveil cases of product shape changes brought about from the influence of technology in three categories given below:

1. Changing the form in a visible way (but not the product as an object ceases to exist);
2. Changes in the product in situations in which technology changes in not reflected in changing the form of the product much but is responsible for modification of performance (performance), keeping the existing product as an object;
3. Cessation of existence of the product as such, in situations where the change in technology leads to a deconstruction of the product as an object, leaving behind the archetypes and stereotypes in a way hitherto associated with the product.

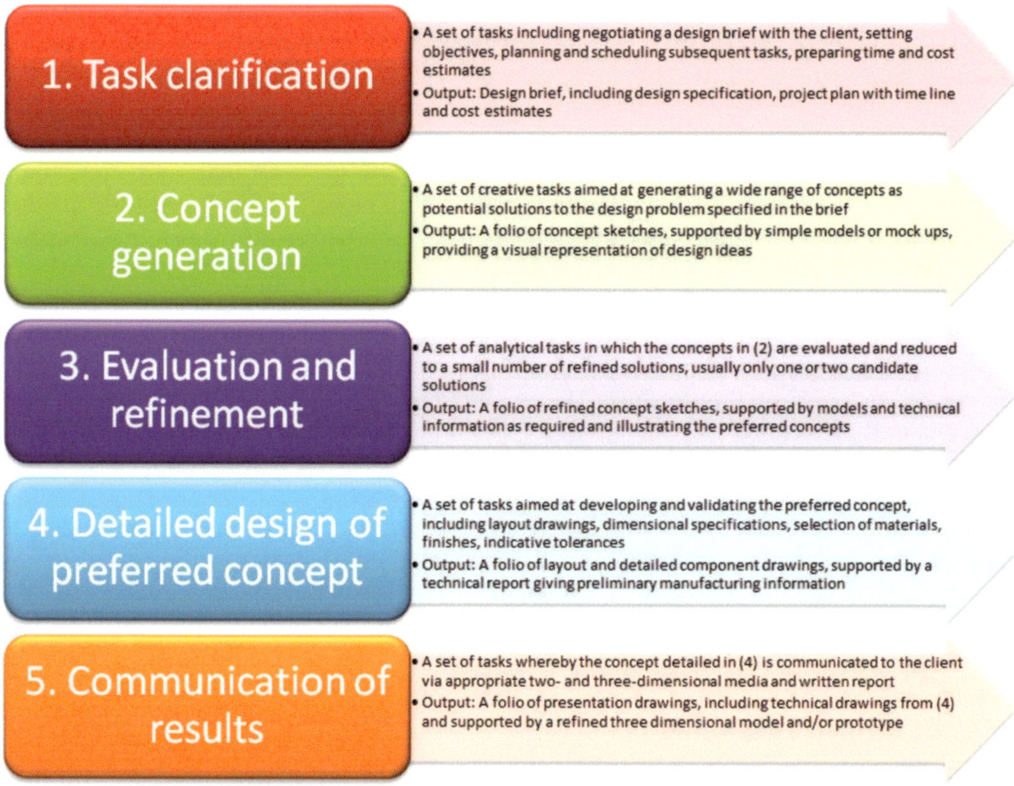

Table 1. Operational Model of the Design Process (Lewis & Bonollo 2002, Hales 1991)

2. Aims

There is an important question that impinges on the future of the practice of industrial design - will there be objects that are at risk of deconstruction, to merge with the environment, in view of the technologies that are developed for the near future and that will replace the existing ones? The overall objective of the work (Corda 2010) reported in this chapter was to carry out a survey of the evolution of form, taking into account the technology of a selected range of consumer technology products in order to deconstruct archetypes that are fixed and propose new concepts, with particular regard to emerging technologies.

With the aim of enabling the widespread use of the methods utilized to answer the aforementioned question and meet the specific objectives underlying the goal of this study, this chapter presents the systematic methodology used so that it can be replicated in other product or technology contexts. This chapter is intended to present the methodologies developed during the study and to explain the course of each of the methodologies and the emergence of a categorization that is divided into three types.

The methodology that is meant to analyse the feasibility of applying an emerging technology in a given product was applied to three cases (Corda 2010), and is made up of five steps in order to analyze and compare the technologies. The intent of this methodology is to attempt to predict if the emergent technology is viable for application in a given product; as such, any technology that is incorporated, or has been incorporated, in a particular product is considered in the analysis and compared with the emerging technology that is intended to come to be incorporated in the same product. As a result of this methodology, one becomes aware of the performance of each selected technology in a given product, and gets to compare the performance of the product depending on the technology that enables the product features (specific and general) that were considered important for the analysis. This methodology is complemented with the use of another methodology, determining the causality of changes in technology on the external shape of the product.

The awareness of the strengths and weaknesses of each technology, in order to verify whether there is any viability in embedding an emerging technology in a product, is followed by the examination of the changes that have occurred by way of deployment of different technologies in the product. The methodology for determining the causality of changes caused by technology (changes in the external shape of the products analysed) also consists of five steps and connects the shape changes occurred in the products with the various technologies that incorporate each particular product.

It is important to provide a comprehensive perspective to support peers who intend to pursue the implementation of the two methods mentioned above, providing a (tripartite) categorization which is primarily a catalogue of the typological consequences of the influence of technology on the form of a product, as a result of applying the two methodologies presented in the process of technological product design.

3. Methodologies developed

This section is intended to communicate to peers (industrial designers and product engineers) the methodological results achieved. These concern stepwise methods to study technologies as determinants of a product's shape and to propose new shapes for products embedding emerging technologies. A number of technologies are considered in relation to a set of products, as a means to show how technology has influenced object shape and how new technologies (e.g. OLED - organic light emitting diodes, MEMS - micro-electro-mechanical systems and energy harvesting) may promote archetype renewal.

Technology is a key player in today's society; it is the engine of our development and our innovation. Predicting its future uses necessitates systematic approaches, attempting to build future scenarios about the way science, technology, society and economy will evolve, in order to promote their benefits and make the most of the impacts that the future may bring (Glenn, Gordon & Florescu 2008 as cited by Damrongchai & Michelson 2009) . According to this report, the look to the future is optimistic as it will bring progress in various fields, including technology, which promises to have the ability to make the world work in a better way than it does today. There are a number of technologies that, according to Bengisu (2003) and Bengisu & Bekhili (2006), will be the most promising technologies for the near future. This selection was established based on an approach that relates the number of scientific publications with the number of patents over the years. Those reports acknowledge the existence of many emerging technologies such as nanotechnology, biotechnology, super-insulating materials and structures, hydrogen storage and combustion technology. OLED, MEMS and Energy harvesting were chosen because they are emerging technologies that already are used and are soon to emerge on a large scale in the market. Additionally, because the technology and product pairs focused in this chapter were chosen simultaneously, the technology and the product had to be mutually compatible, so they could be articulated (technology and product) enabling the creation of new designs.

Organic Light Emitting Diodes (OLED), which are also called Light Emitting Polymers (LEP), are a technology which is at the forefront of bright screens and monitors, and has been steadily developing in recent years. OLEDs reached the media headlines in 2003 with one million units sold as part of a small application for an electric shaver from Philips, which gave an indication of the level of battery charge. Sometime later, a colour screen (OLED) also appeared with great success as a monitor on the back of a camcorder; Kodak (a major driving force in developing this technology) finally launched this image technology for the world market (Salmon 2004).

The development of MEMS, or micro-electromechanical systems, is responsible for an endless number of modern features and applications. This is a technology that has existed for some time but has expanded greatly in recent years, becoming an increasingly promising technology for the future. As we move into the third millennium, the number of applications for MEMS in our daily lives continues to increase, promoting continuously falling production costs for these devices (Beeby et al. 2004).

Energy harvesting (EH) approaches, form a group of means to harvest energy, which due to scarcity of natural resources such as crude oil, and increasing pollution of the planet by some forms of energy production (such as polluting power plants), have began to gain importance and relevance in the production of clean energy. Increasingly there is awareness that every contribution that can save on energy from pollutant sources is welcome. As such, the use of personal devices, with the ability to produce a few milliwatt of power (a thousandth of the electric power needed for a common light bulb), coming from sources captured by forms of micro energy harvesting, are aligned with this purpose.

The first approach that was developed in the study reported in this chapter was to verify the potential application of an emerging technology in a given product, evaluating the application of OLED technology in TV sets. Five specific aspects and five general aspects were selected that were considered crucial to the performance and the quality of this product. In the second case, this methodology was used to predict the feasibility of applying EH technology (which is an approach to self-powering of technological devices, a grouping of forms of energy harvesting) to the clothes pressing warm iron, and likewise five general and five specific aspects were chosen that were considered relevant for this product and that enable the full unfolding of the methodology. The third case of deployment of this methodology consisted in attempting to assess the feasibility of application of MEMS to the vacuum cleaner. Each of the five steps that make up this methodology will be explained in this section, as well as their limitations.

Another methodology is demonstrated in this section. It is intended to assist in determining the causality of changes in technology towards changes in the external shape of products. It hence aims at assessing technology's causality in relation to the changes that have occurred in the shape of the products under study (TV, iron and vacuum cleaner) as technology changed in them, attempting to predict which aspects of shape may arise and disappear with the implementation in these products of the selected emerging technologies. This methodology was applied first to the TV set, and was used to compare the shape in this product, depending on the technology that comprised it. Then the same methodology was employed to try to determine the aspects of form of warm clothes pressing irons according to the technology that they embody, and finally, to determine those aspects of the shape of vacuum cleaners, also depending on technology.
Observation of results and data taken from the application of these two methodologies gave rise to the need to create a comprehensive categorization of all cases of changes occurring in the form of products, which derive directly from the technology they incorporate, resulting in a categorization, consisting of three variations that embrace distinct types of shape changes in products.

3.1 Feasibility analysis of the implementation of an emerging technology in a given product

The methodology for feasibility analysis of the implementation of an emerging technology in a product consists of five steps (Table 2).

1st	2nd	3rd	4th	5th
Selection of an emerging technology. Selection of a product that is adequate to the technology selected. Identifying the technologies used (past and present) in the product selected.	Collection of information concerning the technologies identified in the previous step. Selection of salient aspects for comparison among technologies.	Compilation of the data considered relevant, obtained in the previous step, and subdivision between general and specific aspects.	Comparison among the compiled data and order ranking of technologies according to their viability in the product.	Selection of the technology with the most promising future for the product focused.

Table 2. Stages of the methodology for feasibility analysis, considering the implementation of an emerging technology in a product.

The first step of this methodology is implemented by choosing an emerging technology, which is intended for study, followed by the choice of product for which the test of feasibility of applying this technology is intended. This is followed by the identification of the technologies used and in use in this product, so that one can establish the comparison and proceed with the remaining actions prescribed in the methodology. The collection of information on technologies makes up the second step, in which one should gather as much information as possible, including the advantages and disadvantages of the technologies that the product uses, and even has used as well as the available data on the emerging technology. These enable drawing up a Table split between the advantages and disadvantages depending on the technology (view example in Table 3, developed for a specific technology - plasma, used on TVs). The third step consists in compiling the data that were considered relevant, after the drafting of the previous step, to create a Table containing the technologies selected in the first step, including the emerging technology (view example in Table 4 – developed for the vacuum cleaner and with the MEMS emerging technology in mind). Five general aspects considered important to the product (for which the study of the feasibility of applying the emerging technology is intended) and five specific issues which relate to factors that influence the product's performance are chosen. This Table is populated with a range of attributes ranging from acceptable, satisfactory, good, very good or excellent, to qualitatively describe the suitability of the technologies in the ten areas selected for comparison. The comparison of the data collected and the ranking of technologies in order of feasibility, comprise the fourth step of this methodology, which transforms the scale of words used in the previous step in a numerical scale ranging from one to five, where number one corresponds to acceptable and number five to excellent and the remaining values follow the order of the verbal scale.

Advantages	Disadvantages
•It is much easier and cheaper to be produced in large sizes, because plasma pixels are very large (de Vaan 2004).	•Large energy consumption. PDP TVs do not get along very well with still images and quickly get burned-out (Salmon 2004).
•Best suited for dark environments because of great color fidelity and contrast (de Vaan 2004).	•Becomes very hot. Long life span. Tend to reflect room lights. Large energy consumption (de Vaan 2004).
•No need for backlight. Wide viewing angle. Excellent contrast and perfect blacks. Good image quality (Salmon 2004).	•Plasma pixels flicker, which causes visual discomfort. Luminescence and color saturation deteriorate over time (Tseng, Chen & Peng 2009).

Table 3. Advantages and disadvantages of the PDP (Plasma Display Panel) technology in TV sets.

The fifth step entails drawing up a matrix (an example of such is given in Table 5, concerning technologies pertaining to the clothes ironing product) which sums up the values assigned in the fourth step, which in turn had been assigned by matching words and numbers in step three. General aspects are added, which enables concluding which technology has greater suitability, according to these aspects. Finally, the results of general and specific features are summed up, and it is then estimated which is the technology that is more feasible to apply the product selected in general (of the two kinds of aspects, both general and specific).

Technology -> Product Feature v	Mechanical (manpower)	Electrical (without AI)	Robotic	MEMS ** (estimate)
Airflow (volume of air intake per second) #	Acceptable	Good	Good	Good
Autonomy (independence of external power) #	Acceptable	Low	Excellent	Very good
Capacity of dirt deposit #	Good	Very good	Good	Satisfactory
Lifetime of the product *	Good	Very good	Very good	Good
Mass (weight) *	Acceptable	Good	Very Good	Excellent
Noise Level (in operation) #	Very good	Satisfactory	Very good	Very good
Power consumption*	Satisfactory	Good	Very good	Very good
Price *	Obsolete	Great	Satisfactory	Acceptable
Vacuum pressure (atraction of dirt) #	Acceptable	Very good	Good	Good
Volume occupied *	Acceptable	Good	Excellent	Very good

Legend: * - General characteristics, applicable to technological products; ** - Concerning the envisaged concept presented in Figure 6, # - Specific features of the product family in question.

Table 4. Qualification of characteristics of technologies concerning the vacuum cleaner.

Technology / Feature	Non-electric	Electrical	Energy harvest
Ability to press multiple kinds of fabric #	1	5	3
Availability and convenience of the energy source #	1	4	5
Ease of Use #	1	4.5	4
Energy efficiency (conversion of electricity in heat) *	1.5	4.5	5
Lifetime (of the product) *	4	3	4
Mass (weight) *	1	4	3
Price *	0	2.5	1
Safety #	0	4.5	5
Versatility of shapes, styles and sizes #	2	4	0
Volume occupied *	2.5	4	4
Total valuation	14	40	33
Rank Order	3rd place	1st place	2nd place

Legend: *- General characteristics, applicable to technology products
- Specific to this particular type of product.

Table 5. Classification of technologies in comparison, for the product clothes pressing iron.

Tables 2 to 5 illustrate how the first methodology presented in this chapter unfolds; its first application demonstrated was to test the feasibility of application of OLED technology in televisions. It has proven very efficient and straightforward to conduct the entire process, as the abundance of data is large, which made it easier to collect relevant data, as well as performing the choice between specific and general issues for comparison between technologies. Another advantage, which allowed the development of this methodology, was that the emerging technology (OLED) has to be applied to the product concerned (TV) and as such values and considerations already existed and were already tested and proven for their practical implementation in the product. With regard to further application of this methodology in the case of the OLED technology within TVs, all the technologies selected for comparison are currently in use in this type of product, so this is not a product with a single dominant technology as is mostly the case for irons and vacuum cleaners. In the other two cases of deployment of this methodology, for EH and MEMS technology, the feasibility of implementing these procedures for the products iron and vacuum cleaner, respectively, was verified. Of particular importance was the fact that these technologies do not have widespread commercial application in the chosen products, and as such it became more difficult to collect data and make accurate analyses of the aspects in comparison between technologies. This was offset by the emergence of estimates and by basing the assessments on envisaged concepts developed for these two products with the incorporation of the emerging technologies. Another hardship found was that predecessors of the technologies used in the products iron and vacuum cleaner, are completely obsolete and are not used anymore, which also hampered the collection of some data.
It should be noted that this methodology is deemed suitable to application for most technologies, and serves the purposes of testing the feasibility of applying a particular technology within a product. It is a methodology that can be improved with the emergence of new data and of results of the application of emerging technology in the product focused.

3.2 Determining causality in cases where technology changes the shape of the product

The methodology for determining the causality of changes in technology, changes in the external shape of the products, is composed of five steps (Table 6).

Selection	Observation	Listing	Emphasizing	Systematizing
Selection of product for analysis. Study of the evolution of the product over time. Categorization of product variants according to technologies incorporated.	Observation of a wide range of examples of the product focused. Taking note of the most salient shape features, according to each technology.	Listing the product shape features found for each technology the product may incorporate or has incorporated.	Emphasizing the most significant shape features that occur in the product, according to each technology considered.	Systematically collecting the results of the previous analysis about the influence of technology on the shape of the product.

Table 6. Stages of the methodology for identification of causality of technological change in the shape alterations of a specific product over time.

The first step of this methodology is to choose the target product for the analysis in terms of form, which is followed by a historical study on the evolution of that product shape from its beginnings to the present. An analysis of the product is then made according to the technology that it incorporates, and may incorporate (the emerging technology that has been selected for the product). Then the form search begins, composed of the observation of a wide range of issues, consulting catalogues and product photos, which are used to infer all the salient features of shape and are organized according to the technology incorporated, completing the second step in this methodology. Creating Tables with the characteristics of the product form (which had been collected in the previous step) split according to the technology that the product incorporates, embodies the third step, which gives rise to a series of Tables. The fourth step consists of highlighting the most significant differences that occurred in the product shape according to technological change, in several Tables listing these changes. This does not only concern similarities, as gains and losses of form features are also of interest. The purpose of this step is to provide a systematic overview of the existing changes with the onset of another technology in a product. In the fifth step, the results from the previous step on the influence of technology in the form of a product are gathered in a systematic way. This enables describing the similarities in appearance, and the changes in form, for the product under study, on which a new technology is to be implemented. This step concludes the deployment of the methodology used to assist in determining technology's causality in changes in shape occurring in TVs, clothes pressing irons and vacuum cleaners.

The main limitation observed of the use of this methodology concerns the fact that in the case of irons and vacuum cleaners there is no model that uses the emerging technology that has been associated with each of these products. For television sets, it was not necessary to resort to the verification of the concept developed in the form of a TV with OLED technology, because actual examples already exist of this product incorporating OLED technology. The fact that in most products there is a vast variety of forms, is yet another hardship in the deployment of this methodology, to the extent that it would be impossible to analyze them all. Moreover, the existence of many technologies (mainly non-electric technologies, in the case of irons and vacuum cleaners) would yield very large lists, covering only a small set of products that used these technologies. As a way to overcome this obstacle, all non-electric technologies were considered jointly and a note was made of the

common aspects of the products embedding them. In the case of the vacuum cleaner, non-electric technologies were grouped together under the manpower label, comprising technologies that enhance the functionality of the product, such as tightening mechanisms, levers and mechanical cranks.
This methodology, which is deemed applicable to most types of products and technologies, is intended to convey a process to collect the aspects of form that make up a product depending on the technology it incorporates, and to relate the change of its shape with implementing a specific new technology in this product. Figures 1, 2 and 3, depict sketches of the salient shape contours of the three products analyzed, according to the type of technology that they embedded.

Non electric iron

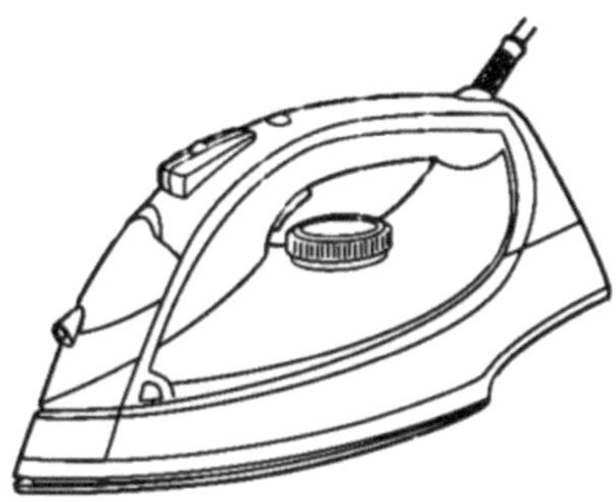

Electric iron

Fig. 1. The evolution of form in the modern clothes pressing iron.

While the fundamental shape archetype is unchanged as the clothes iron passed from non-electric to electric energy (Figure 1), over time there has been a gradual evolution of the form features in this product, resulting in a lighter and more streamlined product. Fundamentally, the product depicted in the right sketch still consists of a V-shaped metal base attached to a handle, and as such, this basic product archetype has endured over time.
The evolution of the shape of the TV set (Figure 2) has clearly been influenced by technology, in particular by the leap from CRT (Cathode Ray Tube) to LCD (Liquid Crystal Display). The most recent technological substitutes for LCD technology (Plasma, LED and OLED) have so far not promoted a fundamental change to the new archetype of the flat and thin TV set. This notwithstanding, there is a clear direction in the evolution of the shape of TV sets towards ultra-thin display panels (and flexible display panels are in the horizon).
Of the three products analysed in this chapter, the vacuum cleaner (Figure 3) is the case where more striking form changes took place as an effect of changing technologies and physical principles that make up its fundamental functionality. In this product, several archetypes of shape have coexisted over time, but clearly, each form archetype is associated to a particular technology, even if the same basic technology underlies several product form archetypes (this is especially true in the case of electric vacuum cleaners, from the pre-artificial intelligence era).

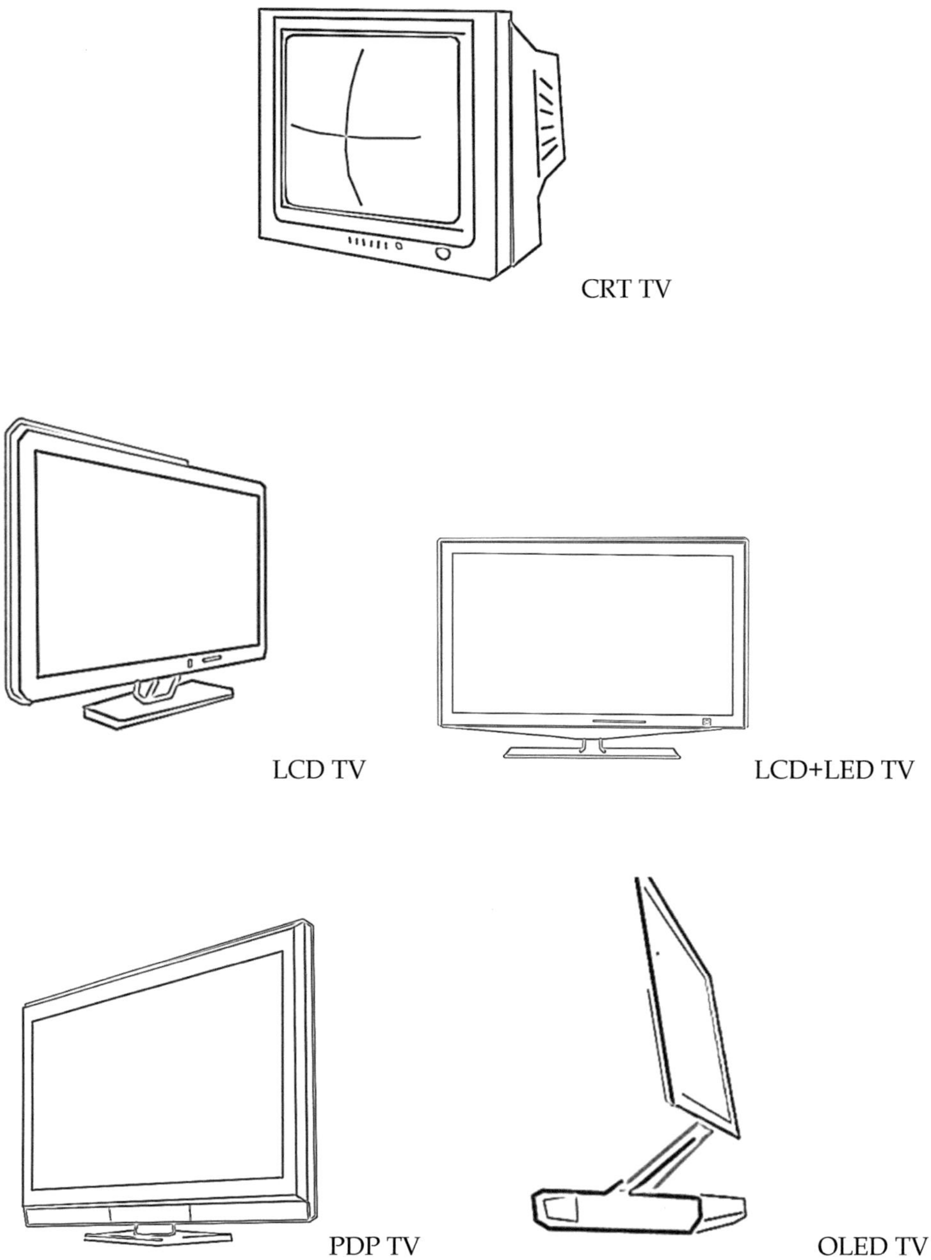

Fig. 2. The evolution of form in modern Television sets.

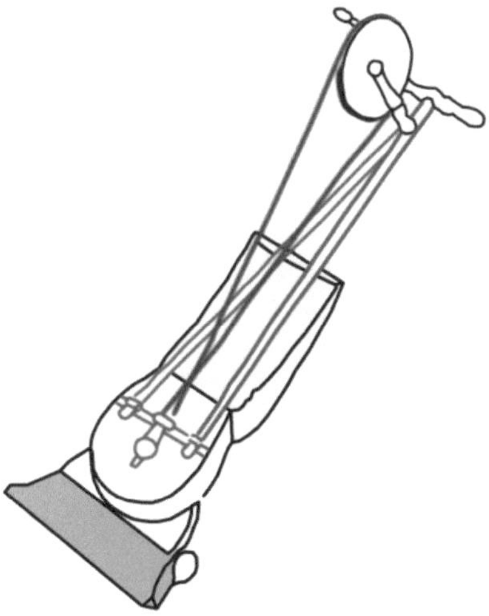

Mechanical predecessor to the vacuum cleaner

Electrical Vacuum cleaner (without artificial intelligence)

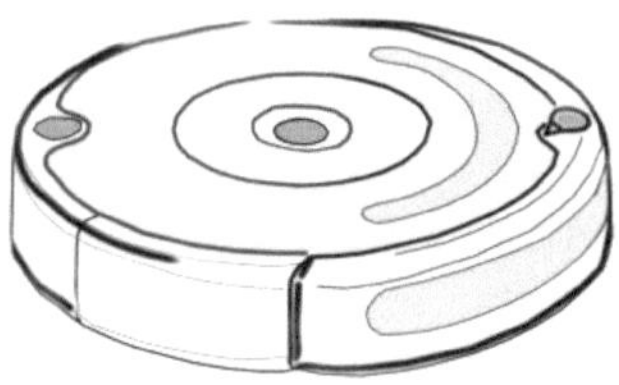

Robotic (automatic) vacuum cleaner

Fig. 3. The evolution of the form of the vacuum cleaner.

3.3 Categorization encompassing three kinds of technology driven shape changes in products

The methodology for determining the causality of changes in technology, changes in the external shape of the products, and the methodology for the analysis of feasibility of application of an emerging technology in a product, encompass the collection of data on the characteristics and performance of products according to the technology that they embody, as well as of the data on the morphological differences occurred due to that technology change. Encompassing the various kinds of form changes occurring in technology products,

necessitates creating a categorization that considers the many changes in technology products. This is a tripartite categorization, which is divided into three variations.
This categorization served throughout the study reported in this chapter to frame the kind of change that occurred in the product after another technology incorporated it. Thus, changes were classified according to variations that occurred during the passage of television sets by various technologies (CRT, LCD, LCD + LED, PDP and OLED), as well as the shape changes resulting from the passage of irons by various technologies (non-electric, electric power and the principle of energy harvesting - EH), and, finally the changes in the vacuum cleaner with the emergence of various technologies (mechanical - human strength, electrical - without Artificial Intelligence, robotics and MEMS).
Thus it is possible to characterize the changes that occur in the products that were studied, within the following three categories:
1st type - change in the shape of the product caused by changing technology (appearance of a new technology or application of an existing technology but that was never used in this type of product) which leads to a visible shift in the product shape, yet the product as an object remains;
2nd type - product change, in situations where technology change is not reflected so much in changing the shape of the product, but is responsible for modification of performance and improved efficiency, keeping the product as an existing object and proceeding to surface shape change in order to signal the increased performance to consumers, and, finally, 3rd type - cessation of existence of the product as such, in situations where the change in technology leads to a deconstruction of the product as an object, leaving behind the archetypes and stereotypes hitherto associated with the product.

4. Conclusion

The aim of this chapter was to undertake a study about the influence of technology in the form of products that incorporate technology, using a standard methodology developed for this purpose. Examples of products and their shape changes over time were presented. Attempts to understand the extent to which these changes are consequences of the developments in technology were put forward, and the importance of technology as a determinant of the shape of products that incorporate technology was demonstrated based on three cases.
A review of the evolution of form, taking into account the technology of a selected range of consumer products that incorporate technology, with the aim of deconstructing archetypes that are fixed and propose new ideas was accomplished. In particular, a range of technologies that are foreseen for the future were considered, and these were also the subject of study in this contribution, and are expected to enable new designs and future forms for a great variety of products, which in some cases may be subject to a deconstruction of the archetype of the object's shape.
Selecting specific cases of products, from a historical perspective, one could see what have been the determinants of their shape, which led to the establishment of their archetypal form. A comparative methodology was developed, in order to hint at the role of technology as a determinant of the shape of products.
Understanding the importance of technology as a modelling influence for products' form, what role this has in the consumer society and what is its responsibility as a major element

in the deconstruction of the archetypes of form was also focused in this contribution. The influence of technology in the transformation of products was demonstrated, considering three distinct types of alterations caused by changes in technology, in a tripartite view.

Following the study of the technologies, which was methodologically structured, and concerned the technologies embedded in the three products in analysis, in the past and with a forward view (technology that is foreseen for the future), a scenario is proposed, with new designs for the products that have been studied (Figures 4, 5 and 6). It is emphasized that the product as a physical object may disappear or may be dissolved in the environment or building architecture, with the advent of emerging technologies and their new capabilities.

Fig. 4. Multiple applications of OLED technology in a living room (lighting, shading, dynamic wall art gallery, TV monitor).

In the case presented in Figure 4, dematerialization of a product (TV set) and its blending with architecture is enabled by the ultra-thin OLED technology. This technology also enables new applications, including, as depicted, lighting, shading and a dynamic wall art gallery. This case leads to question the role of industrial design in this foreseen evolution. Understanding people and their relation to artefacts, interaction design and concept creation (even if devoid of a three dimensional form) are bound to be design domains that industrial designers should focus on in supporting this kind of design endeavour.

The conceptual design presented for an energy self-sufficient travel iron (Figure 5), represents an archetypal leap from the traditional shape of this product, which caters to sustainability concerns. Independence of energy supply is bound to be another guiding theme to foster the creation by industrial designers of new product archetypes to perform functionality that had been tied to a fixed product concept.

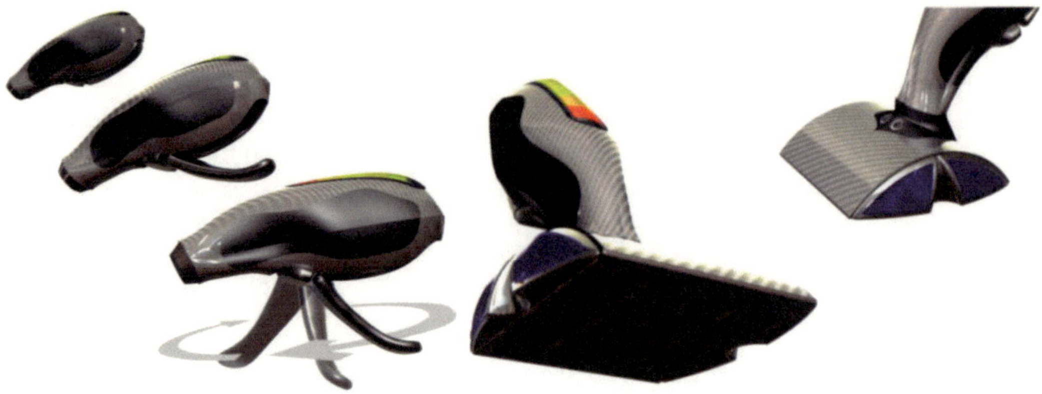

Fig. 5. Energy self-sufficient travel iron.

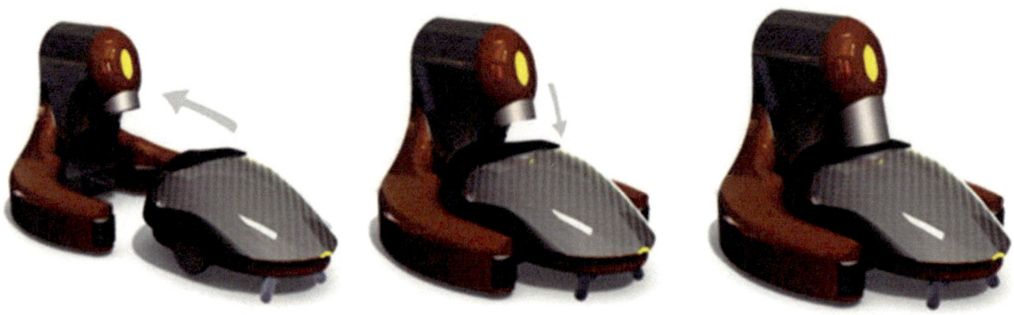

Fig. 6. System for unloading and battery charge of robotic vacuum cleaner.

The conceptual design presented in Figure 6 represents an incremental shape evolution from one of the current shape archetypes for vacuum cleaners. As growing automation and convenience for consumers proceeds, industrial designers are also bound to play an important role in establishing links with past era designs, and as such provide consumers in this digital age, objects that enable reminiscing of past eras, with a shape that cues their functionality, providing a deliberate a link with traditional shapes.

There are limitations in the methodologies presented in this chapter, which will be alleviated if the prerequisites that are necessary for their effective use are met. The methodologies and categorization are only deemed suitable for application in products that incorporate technology for their operation. It is important for the deployment of these methodologies to have knowledge on emerging technologies and on the history of the product or devices that have been used for implementation of a particular functionality.

There are some historical cases of technologies that have failed, and that were put aside, because there was an arrest of the market by a technology already implemented. This happened for example with Sony's Beta video system, which was of higher quality than Philip's VHS (Video Home System), but the latter prevailed (despite the better quality of the Beta video system), because films existed in greater quantities in the VHS format, which led to the decrease of availability of films in Beta format. There is no guarantee that the three emerging technologies selected will have a future in the market, or even a future in the product (or application) that has been studied. The fact that emerging technologies are still developing, makes their performance characteristics modifiable and changeable in a short time span. It is also noteworthy that the current concerns about environmental sustainability have led governments to constrain the free operation of markets which can alter the dynamics of competition for emerging technologies, seeking to accelerate the implementation of more sustainable alternatives, and restricting the widespread adoption of other technologies, which may discourage the implementation of some emerging technologies.

5. References

Baudrillard, J. (1995). Sociedade de consumo [in Portuguese - The consumer society], Lisboa: edições 70.

Beeby, S., Ensell, G., Kraft, M. & White, N. (2004). MEMS Mechanical Sensors, Southampton, United Kingdom: Artech House.

Bengisu, M. (2003). Critical and emerging technologies in materials, manufacturing, and industrial engineering: a study for priority setting, Scientometrics, Vol. 58, No. 3, 473-487.

Bengisu, M. & Bekhili, R. (2006). Forecasting emerging technologies with the aid of science and technology databases, Technological Forecasting & Social Change, Vol.73, 835-844.

Bloch, Peter H. (1995). Seeking the ideal form: product design and consumer response, Journal of Marketing, 59(3), 16-29.

Corda, F. A. A. (2010). A tecnologia como determinante da forma dos objectos [in Portuguese - Technology as a determinant of object shape], Master of Science Dissertation in Technological industrial Design, Dept. Electromechanical Engineering, Universidade da Beira Interior, Portugal. Available on-line at http://webx.ubi.pt/~denis/MDIT/dissertacao_FilipeCorda.pdf

Crilly, N., Moultrie, J. & Clarkson, P.J. (2009). Shaping things: intended consumer response and the other determinants of product form, Design Studies, 30(3), 224-254.

Damrongchai, N. & Michelson, S.E. (2009). The Future of Science and Technology and pro-poor applications, Foresight, 11(4), 51-65.

Glenn, J.C., Gordon, T.J. & Florescu, E. (2008). 2008 State of the Future. The Millennium Project, Washington, DC.

Hales, C., (1991). Analysis of the Engineering Design Process in an Industrial Context, Eastleigh, UK: Gants Hill Publications.

Lewis, W. P., Bonollo, E., (2002), An analysis of professional skills in design: implications for education and Research. Design Studies No.23, pp. 385-406.

Roozenburg, N. F. M. & Eekels, J., (1995). Product Design: Fundamentals and Methods, Chichester: John Wiley & Sons.

Salmon, R. (2004). The changing world of TV display - CRT challenged by flat-panel display, EBU technical review, April 2004, pp.1-9.

Tseng, F., Cheng, A., & Peng, Y. (2009). Assessing market penetration combining scenario analysis, Delphi, and the technological substitution model: The case of the OLED TV market, Technological Forecasting & Social Change, Vol.76, 897–909.

de Vaan, A. J. S. M. (2004). Competing display technologies for the best image performance, Journal of the SID, 15, 657-666.

2

Knowledge-Based Engineering Supporting Die Face Design of Automotive Panels

Chun-Fong You, Yu-Hsuan Yang and Da-Kun Wang
Department of Mechanical Engineering, National Taiwan University
Taiwan, ROC

1. Introduction

Sheet metal is commonly used in the automotive industry. The requirements for developing approaches for manufacturing automotive panels have become increasingly important as carmakers seek to shorten their time to market. However, designing and manufacturing mold dies for automotive sheet metal panels is time-consuming as several processes use the trial-and-error method. The die design period for one sheet metal panel is usually a few months to up to one year.

Along with advances in Computer-Aided Design (CAD), simulation of interference checking can be conducted using a three-dimensional solid model, and Computer-Aided Engineering (CAE) software can be used to simulate and analyze die face to accelerate the design process (Fig. 1).

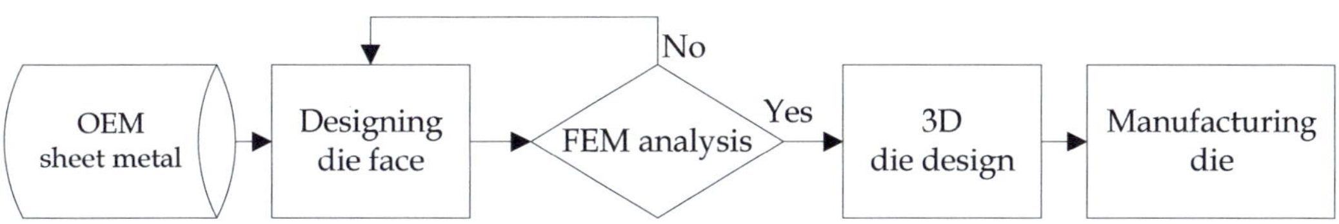

Fig. 1. Design process of die design

Cold dies are used for drawing, trimming, restriking and piercing when manufacturing sheet metal (Fig. 2).

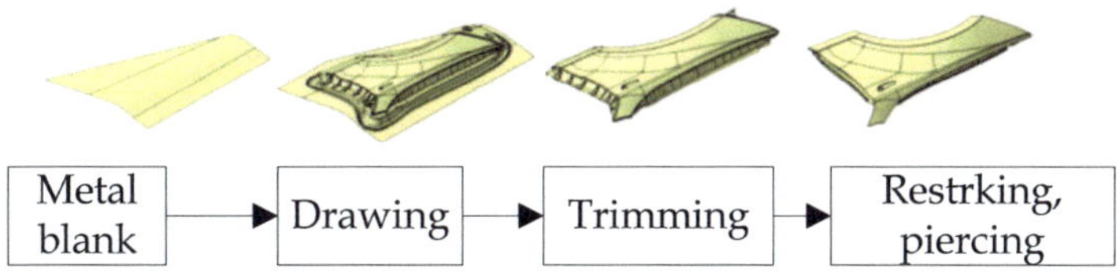

Fig. 2. Manufacturing process of automobile sheet metal

1. Drawing (DR) mold

Drawing is the first operation in a manufacturing process in which a metal blank is drawn without wrinkles or cracks.

2. Trimming (TR) mold

A trimming mold is used to cut sheet metal into an appropriate size and shape, such that the restriking mold can bend. During this operation, the most essential factor is trim cutter and scrap cutter position in a combination that enables sheet metal scrap to automatically drop and be removed after trimming.

3. Restriking (RST) mold

Restriking is an operation in which sheet metal is molded into a desired shape. The main activities in this process are flanging and restriking. Flanging refers to the process of folding edges at a 90° angle, while restriking refers to folding at any angle and is usually driven by a cam for wide areas. When designing a mold for restriking, one should compensate for springback, which can be determined based on raw material characteristics.

4. Piercing (PI) mold

Piercing is typically the last operation. When piercing is done prior to restriking, the position of a hole can move or hole shape can deform during shaping.

2. Die face

The die face plays a crucial role in drawing operations as it determines operation quality (Makinouchi, 1996). Not only does the design have to prevent sheet metal from cracking, wrinkling, offset of the characteristic line, it also has to take the following factors into consideration including compensation of springback, and whether the trimming mold should use normal cams or suspended cams, whether the restriking mold should use a convex hull, and piercing direction. Designers usually search for previous successful cases as references when designing a new die face. This design process is explained in Section 2.2.

2.1 Introduction to the die face

Fig. 3 lists the factors that should be considered when designing a die face for an automotive fender.

1. Product-in face: The part of a sheet metal that would be shaped during the drawing operation (excluding the areas that would be folded afterwards).
2. Binder: The binder is the area that a piston presses against on an upper mold to ensure the metal blank remains stationary during drawing.
3. Drawbead: The drawbead controls tension resistance of a metal blank during drawing.
4. Convex hull: The convex hull stores prepared materials to avoid sheet metal cracking while folding.
5. Stamp mark: A stamp mark is used when determining whether the upper mold and lower mold are in full contact.
6. Parting line: The parting line is a line separating the lower mold and piston.
7. Trimming line: The trimming line is a rough profile of the boundary of a piece of sheet metal. During trimming, the sheet metal is trimmed along this line to facilitate restriking.

Die face design can be divided into three parts—product-in face, product-out face, and binder. Fig. 4 shows a die face cross section. A sheet metal surface generally has two parts—product-in face and product-out face. Product-out face is the largest boundary that must be restriked because it can only be shaped after drawing, and is not included in the die face.

Product-in face is the outside of a sheet metal piece that can be seen after being assembled in an automobile and is shaped during drawing and, thus, is included in the die face.

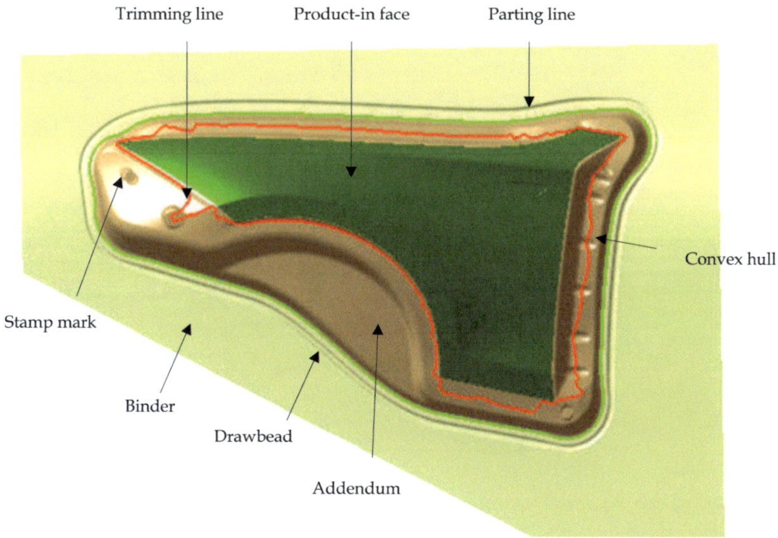

Fig. 3. Features on fender design

Addendum is the surplus area outside the product-in face that facilitates drawing operations; thus, when designing an addendum, the quality and strength of a sheet metal piece after drawing must be considered.

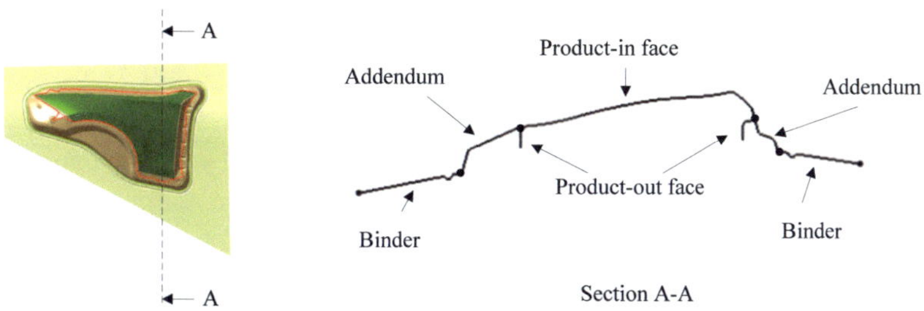

Fig. 4. Divisions on die face

Normally when designing a die face, except for such parts as product-in face, the addendum, and binder—the three main parts—other features like drawbead and the stamp mark during the drawing operation, trimming line during the trimming operation, and convex hull during the restriking operation are also included in die face to facilitate those operations.

2.2 Designing a die face

Fig. 5 shows the design process for a die face. When stamping a piece of sheet metal, no area should be unable to be pressed; that is, undercut (Fig. 6). The upper mold and lower mold should be fully pressed against each other during drawing and all areas that must be drawn should be drawn at one time; therefore, avoiding an undercut is the first priority when designing a die face.

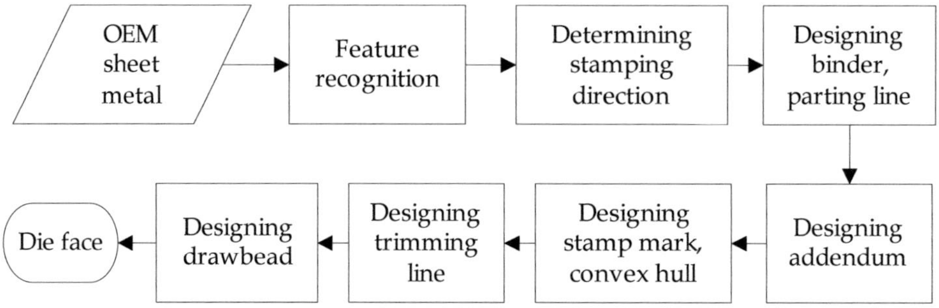

Fig. 5. Design process of die face

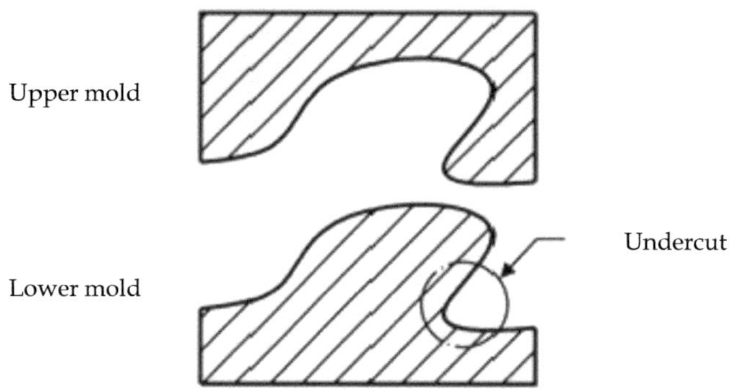

Fig. 6. Undercut

2.2.1 Feature recognition

The first step in designing molds for sheet metal automotive panels is to determine the stamping angle, such that subsequent operations can be successful. Typically, the outer appearance of a panel should be included in the product-in face and be formed during the first operation to yield the highest surface quality with the largest stamping force among all the operations.

A die face can be divided into two parts. One is product-in face and, when designing it, its face cannot have an undercut, and radius of chamfer should be >3mm to prevent cracking while drawing. If these requirements cannot be met, an addendum can be added, such that some tasks can be done in later operations.

The area of a sheet metal outside of a product-in face is called product-out face, which is divided into connecting features and corners (Fig. 7). The design of a product-out face focuses on how to facilitate restriking and bending operations. A product-out face can be divided into several parts and cams can be used to shape each part.

The case in this study considers product-out face as a feature (Fig. 8), and is adopted from the previous works (Tor et al., 2003; Zheng & Wang, 2007).

Factors are considered when designing a die face and cams are the existence of an undercut, area and length of the line connecting the product-out face to the product-in face, and the angle between the product-in face and the product-out face. Thus, this study uses these features to describe the product-out face. In some cases, undercut surfaces may be blocked by the product-in face during stamping (Fig. 9). In such cases, cams should be utilized to change the direction of stamping forces to horizontal to form surfaces that are undercut.

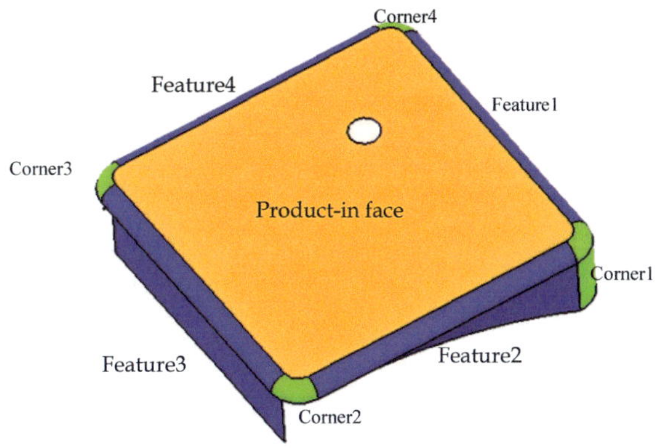

Fig. 7. Surface features on sheet metal

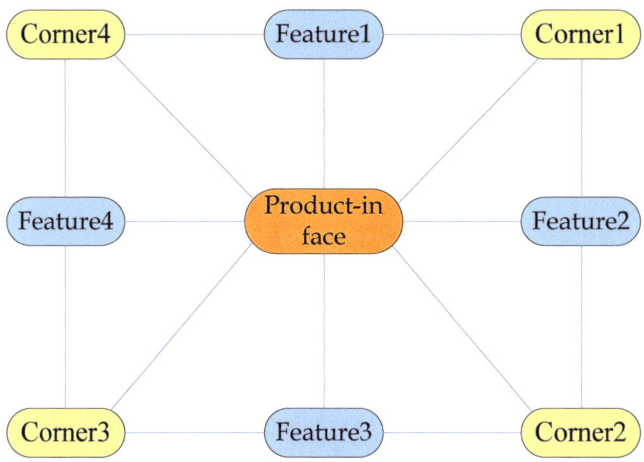

Fig. 8. Feature graph of sheet metal

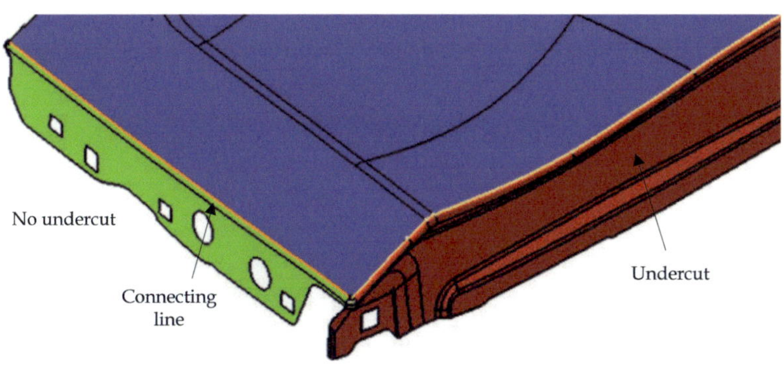

Fig. 9. Features on product-out face

2.2.2 Analysis of stamping direction

The most important drawing goal is to draw a raw metal piece into a desired high-quality shape. This quality is markedly affected by the orientation of stamping when the sheet metal panel is punched and is based on symmetry, equal-angle, and equal-depth not exceeding an appropriate value.

2.2.3 Binder design

To prevent sliding and wrinkling during drawing operations, a binder is used to hold the piston and upper mold. A binder should be the same height as the product-out face and be smooth and simple in terms of geometry. A binder has a straight line, a curved section, and different types of boundaries.

2.2.4 Addendum design

The shape of a sheet metal automobile panel is typically complex and irregular, resulting in difficulty achieving uniform forces on the die face during drawing operations. To solve this problem, an addendum is introduced that uses various section curves (Fig. 10) to make forces uniform. Additionally, the convenience of subsequent operations is also a concern and can affect the choice of section type.

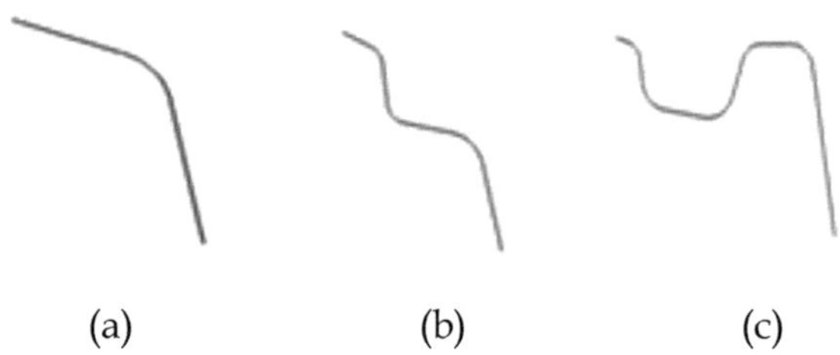

Fig. 10. Common section types

The design of an addendum requires determination of the trimming position, section type, and its size. Fig. 11 provides a detailed explanation of Fig. 10(b). A 3–5mm line is usually extended from the product-in face to avoid cracks from bending. An addendum is typically shaped like stairs to facilitate trimming. The bending angle of an addendum should be as small as possible.

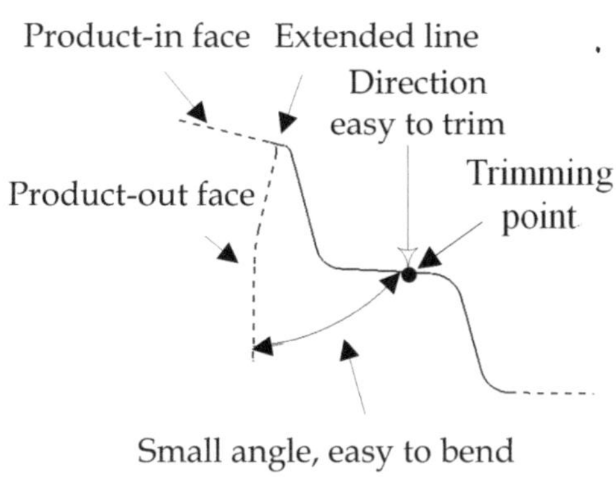

Fig. 11. Stair-like section curve

2.2.5 Addendum construction

Designers first construct the parting line as a limitation in subsequent design steps (Fig. 12). The addendum surface is then designed, which is composed of a section curve and connecting curve (Dy et al., 2008). The section curve determines how the addendum is shaped and is a concern for subsequent operations. The connecting curve connects section curves to produce a smooth surface. Finally, chamfer at the parting line.

When using programs to construct an addendum, the trimming point should lie on the trimming face and a connected curve along the shape of the sheet metal panel should be used to make its surface smooth. The design platform in this study is based on the SpringSolid system developed by the Solid Model Laboratory, National Taiwan University and written in Java.

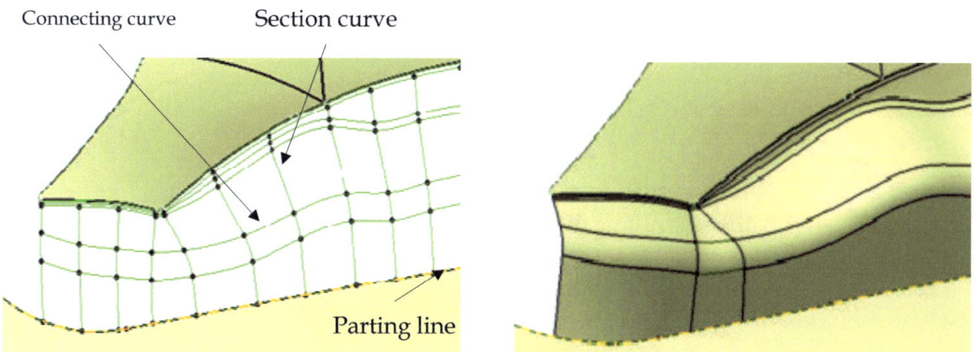

Fig. 12. Addendum construction

3. Knowledge-based engineering

Knowledge-based engineering (KBE) refers to the concept of a knowledge database applied in engineering that can be regarded as an intelligent system in a specific engineering field in which experts modularize product information and design processes to assist in product design. The design process is then stored for knowledge management. KBE is also combined widely with CAD/CAE/Computer-Aided Manufacturing (CAM) software for design, analysis, and manufacturing, respectively.

The KBE system is composed of a database and reasoning engine; the database stores related knowledge and assists in design via the reasoning engine.

Retrieval and case representation of knowledge are two crucial elements of a knowledge database. First, KBE engineers retrieve related knowledge from books, experts, and other resources, and record this knowledge using an appropriate knowledge representation. A representation should be able to store related knowledge in that field to enable a system to read and show that knowledge such that the knowledge can be provided to the reasoning engine.

3.1 Case-based reasoning

Fig. 13 shows case-based reasoning (CBR) operations. CBR compares cases in an analogue way. First, CBR compares a new case with cases in a case database and searches for the most similar case.

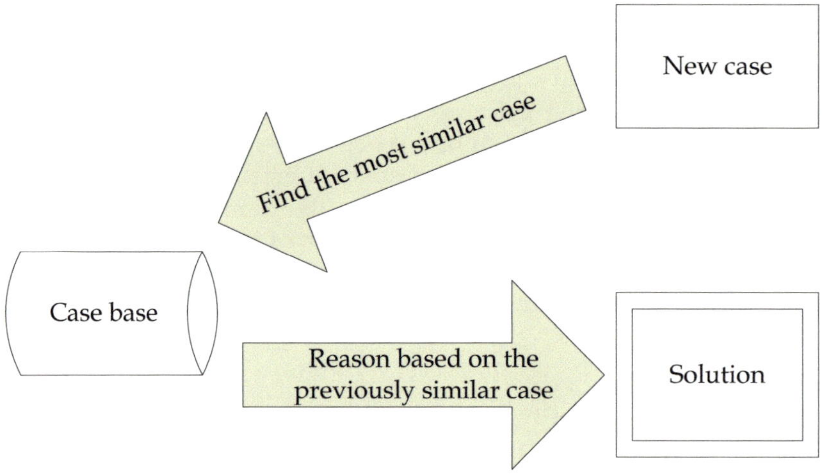

Fig. 13. Reasoning process of case-based reasoning

Typically, CBR has the following four procedures (4R) (Fig. 14):

1. Retrieve: After feature recognition of an automotive sheet metal panel, CBR compares the features with those in the case database, assesses the similarity among cases, and retrieves the most similar case as a reference for the design process.
2. Reuse: Designers can decide whether a retrieved case is appropriate for reuse and which manufacturing method should be a reference.

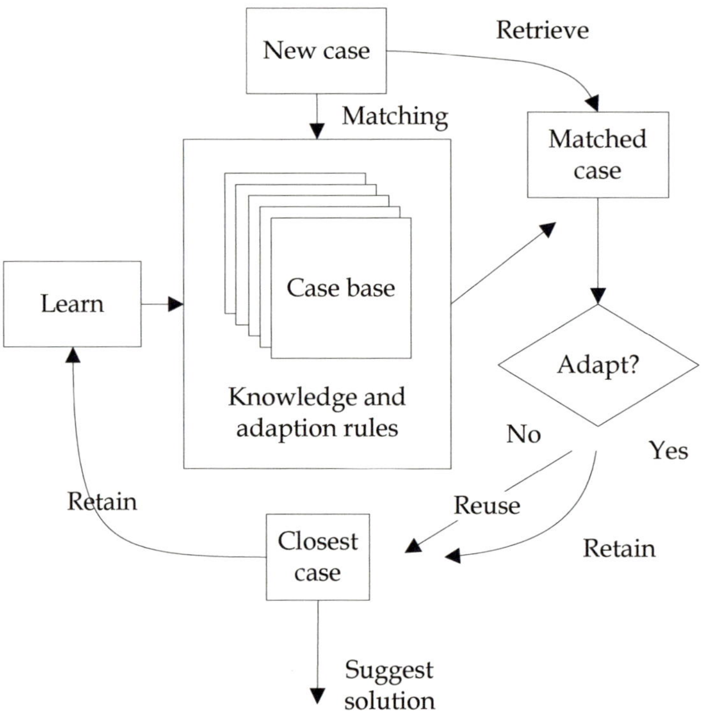

Fig. 14. The CBR cycle (Kendal & Creen, 2007)

3. Revise: Based on the reuse assessment, designers revise the proposed solution when necessary.
4. Retain: The final assessment result for a stamping die design for an automotive sheet metal part for future reference is stored and reference cases are recorded.

Differing from rule-based reasoning, CBR does not require a set of explicitly defined mathematical models, rules, or logic. Thus, CBR is suitable for problems with general rules that cannot be systemized.

When comparing cases, algorithms (Watson & Marir, 1994; Tor et al., 2003) are used to assess similarity among cases, and CBR uses significant features to describe a case (Fig. 15), compares each case with given weights, and finally determines total similarity.

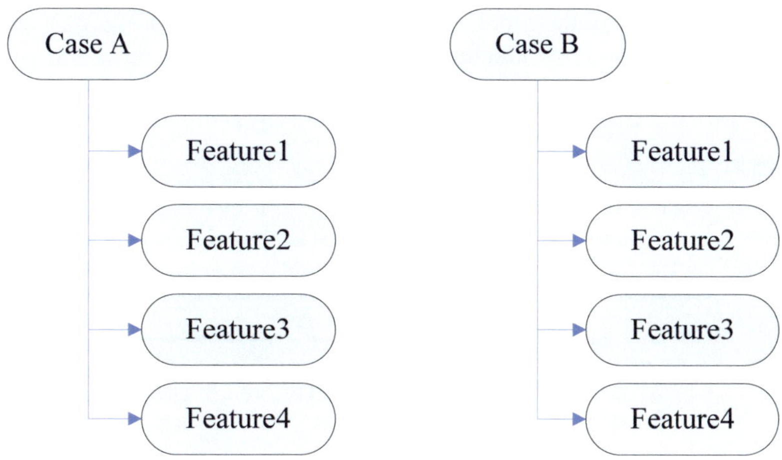

Fig. 15. Data structure of feature recognition

The algorithm for determines similarity is $S = \dfrac{\sum_{i=1}^{n} w_i \times s(f_i^A, f_i^B)}{\sum_{i=1}^{n} w_i}$.

S : similarity between case A and case B, $0 \le S \le 1$

w_i : weight of each feature

f_i^A : the i_{th} feature of case A

$s(f_i^A, f_i^B)$: similarity between features f_i^A and f_i^B, $0 \le s(f_i^A, f_i^B) \le 1$

When features are represented numerically, the similarity between features f_i^A and f_i^B would be $s(f_i^A, f_i^B) = 1 - \left| \dfrac{f_i^A - f_i^B}{\max(f_i^A - f_i^B)} \right|$.

When features are represented non-numerically (e.g., Boolean value or textual description), similarity between feature f_i^A and f_i^B is $s(f_i^A, f_i^B) = 1$ if $f_i^A = f_i^B$; otherwise, $s(f_i^A, f_i^B) = 0$ if $f_i^A \ne f_i^B$.

3.2 Determining similarity of sheet metal panels

In this study, KBE and CBR are combined to provide designers with guidance from similar panels when designing a new panel.

Before comparisons, one should first define "similar" for two sheet metal parts. One approach (Tor et al., 2003) is to use part features, geometries, topologies, and materials to describe panels, meaning that this method compares the "appearance" of sheet metal parts.
However, in this study, locating sheet metal panels that are similar in terms of manufacturing processes is more important than locating those with similar appearances. This is because sheet metal parts that have a similar appearance may be made with different manufacturing processes and, on the other hand, sheet metal parts that have different appearances may have similar manufacturing processes. For instance, two significantly different product-in face parts may have been shaped by the same drawing operation, meaning these differences do not guarantee differences in the manufacturing process.
When comparing two sheet metal parts, the product-in face part should not be considered because it does not significantly affect manufacturing processes. However, the product-out face part significantly affects manufacturing processes. Thus, this study compares product-out face sheet metal parts to locate cases with similar manufacturing processes.
The parts are compared using a cross-reference method to calculate similarity.

1. Cross reference

Each sheet metal part has uncertain number of product-out face areas. Take sheet metal parts A and B as an example; this study first cross-references each product-out face part (Fig. 16).

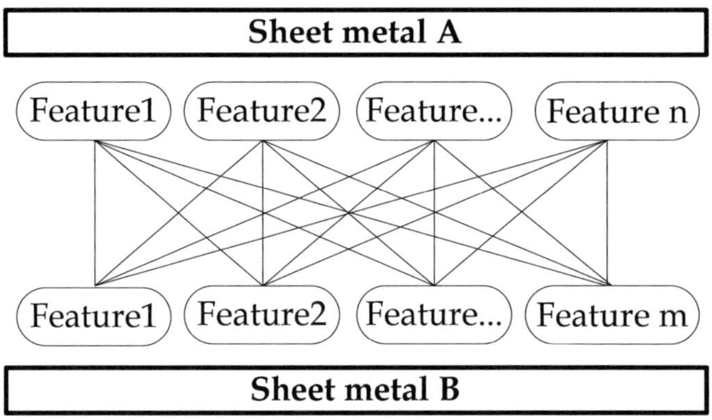

Fig. 16. Cross-reference between sheet metal A and B

Here, S_{ij} is defined as the similarity between the i_{th} product-out face area of panel A and the j_{th} product-out face area of panel B and $\begin{cases} 1 \le i \le n \\ 1 \le j \le m \end{cases}$. An n x m matrix can be derived as

$$\begin{bmatrix} S_{11} & S_{12} & \cdots & S_{1m} \\ S_{21} & S_{22} & \cdots & S_{2m} \\ \vdots & \vdots & \ddots & \vdots \\ S_{n1} & S_{n2} & \cdots & S_{nm} \end{bmatrix}.$$

The maximum entry S_{ab} is then found and marked as S_1 and then $S_{a1}, S_{a2}, \ldots, S_{am}$ and $S_{1b}, S_{2b}, \ldots, S_{nb}$ are removed.

Next, the maximum entry S_{cd} among the n-1 pieces of product-out face parts of each sheet metal is found and marked as S_2; the entries that are similar to the c_{th} product-out face of A and those that compared with the d_{th} product-out face of B are removed. Eventually, each feature of product-out face A that matched those of product-out face B are found (Fig. 17)

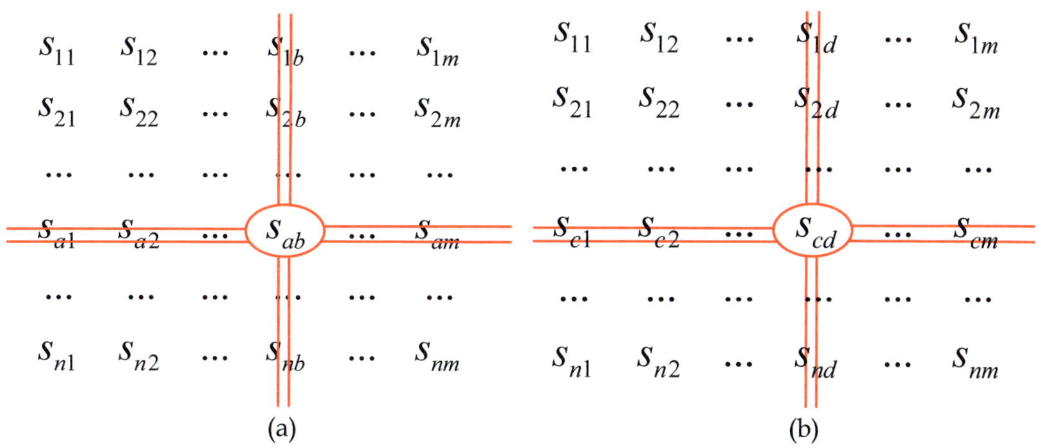

Fig. 17. Accessing similarities between features of A and that of B

This comparison is repeated, such that, $S_3, S_4, \ldots, S_m$ $(n \geq m)$ can be derived and, finally, the similarity between sheet metal A and B can be determined as $S = \frac{1}{n}\sum_{k-1}^{m} S_k$.

In this approach, when the numbers of product-out face areas differ $(n > m)$ between sheet metal parts A and B, the maximum similarity is m/n. As the difference between n and m increases, the similarity between parts A and B decreases. If A and B have the same number of product-out face areas, maximum similarity is 1.

2. Similarities between product-out face areas

Table 1 shows the similarities between product-out face areas. In this table, f_1 is a non-numerical item representing the existence of undercuts. If both sheet metals parts have an undercut, then $s(f_i^A, f_i^B) = 1$; otherwise, $s(f_i^A, f_i^B) = 0$. Items $f_2, f_3, \ldots, f_6$ are numerical ones and their similarity is defined as $s(f_i^A, f_i^B) = 1 - \left| \frac{f_i^A - f_i^B}{\max(f_i^A - f_i^B)} \right|$. If one calculates the similarity between each item with weights, overall similarity between two sheet metal parts is $S = \frac{\sum_{i=1}^{n} w_i \times s(f_i^A, f_i^B)}{\sum_{i=1}^{n} w_i}$.

Similarity	Compared item	Data Type	Weight
$s(f_1^A, f_1^B)$	f_1 : the existence of undercut	Boolean	3
$s(f_2^A, f_2^B)$	f_2 : area ratio (product-out face to product-in face)	float	1
$s(f_3^A, f_3^B)$	f_3 : connecting line (the line that connect product-out face and product-in face)	float	1
$s(f_4^A, f_4^B)$	f_4 : angle (between product-out face and product-in face)	float	1

Table 1. Similarity of items of product-out face

3.3 Application of designing with case-based reasoning

Designing a die face is an art; that is, it is difficult to systemize. However, some sheet metal parts share common features, enabling reuse of similar cases to reduce design time.

Among all procedures when designing a die face, the most difficult task is designing the addendum part, which affects all operations and involves choosing an appropriate section type and size, and determining the position of trim points. All these tasks require tacit knowledge (Polanyi, 1958) and, thus, design time is considerable. Therefore, this study applies CBR to locate a similar case (Schenk & Hillmann, 2004) to accelerate design time.

This study combines CBR when designing a die face for a sheet metal panel with KBE (Fig. 18). To make the system flexible, the scale of reuse can be based on the degree of similarity between two sheet metal parts. If only a few features are shared, then only those features would be adopted; this is called local reuse. Conversely, if many similarities exist, then reuse can involve the entire addendum design; this is called global reuse.

The area or items that can be reused from previous cases are those that are difficult to design and their design is time-consuming. For a die face, only the addendum meets this criterion. The product-in face and product-out face parts are relatively easy to design and are not reused.

Two examples are used to demonstrate how local reuse and global reuse operate.

Fender B is an example of local reuse. After constructing its binder and the parting line of the die face, CBR is applied and the design of the addendum of, say, fender A is used to reduce design time. Due to the shape complexity of fender B, similarity is only for relatively small parts and, thus, only a small portion of the previous design is reused.

The procedures for reusing the design of fender A for fender B are listed as follows.

Step 1. Locate the most similar case—fender A (Fig. 19).

Step 2. Select a portion of the boundary of fender B that is similar to that of fender A (Fig. 20).

Step 3. Select the corresponding boundary of fender A (Fig. 19) and then apply it to the design of fender B. In this case, only the radius of chamfer, draft angle, and addendum design are reused. Other parts, such as the trimming line, are designed all over again.

Step 4. After reusing the design of fender A, a user can decide whether to adopt the reused design. The final reuse result is stored in a database for future reuse.

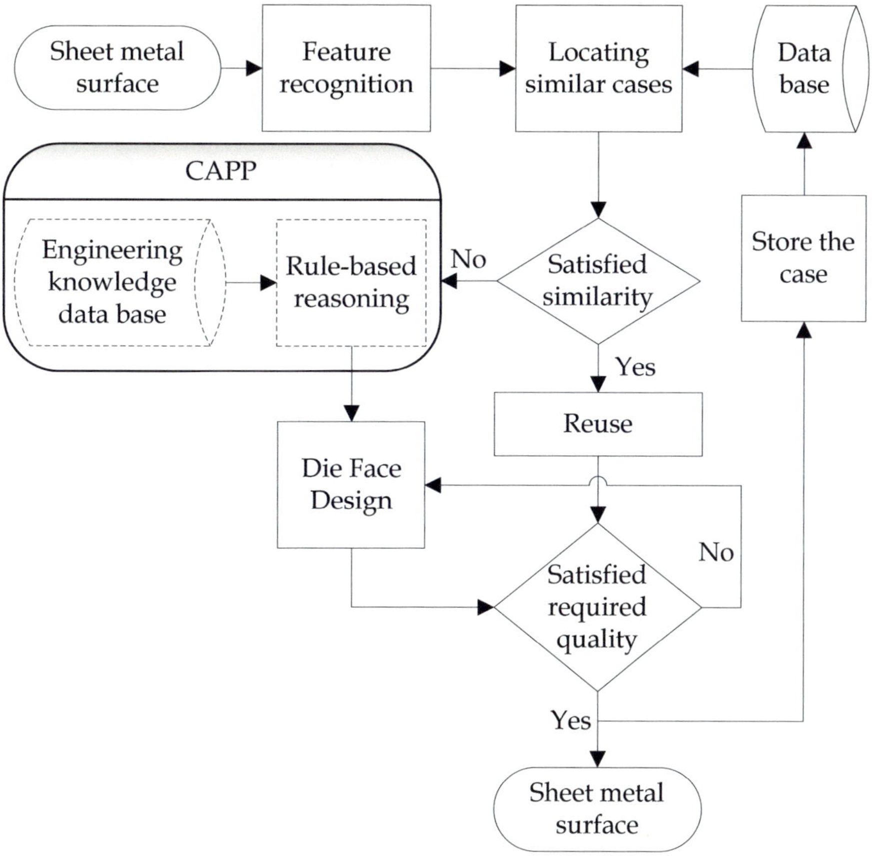

Fig. 18. Design process integrated with case-based reasoning

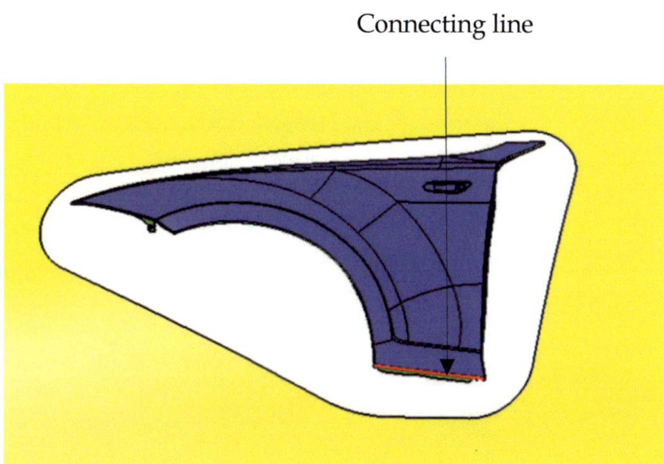

Fig. 19. Fender A

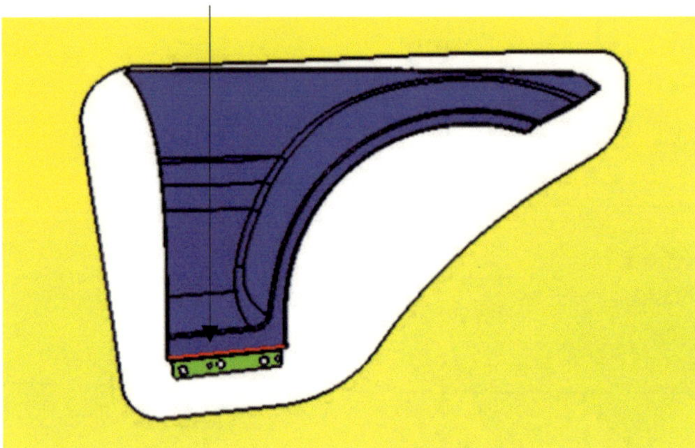

Fig. 20. Fender B

Fig. 21 shows the detailed explanation of addendum A. The reused parameters (Fig. 22) include the radius of section curve and the draft angle (Fig. 23); other parameters of, such as the addendum and trimming line, are new designs.

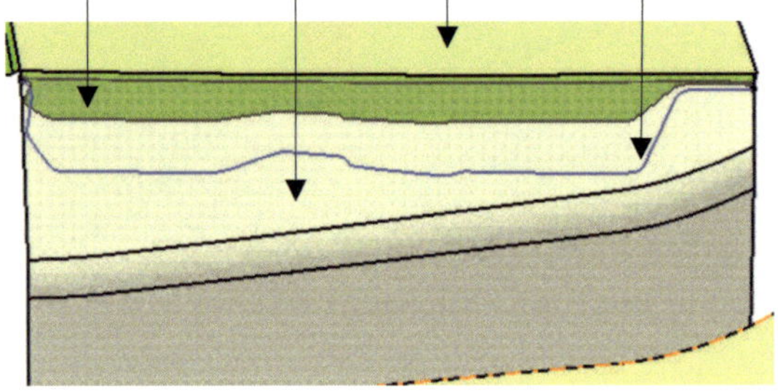

Fig. 21. Addendum of fender A

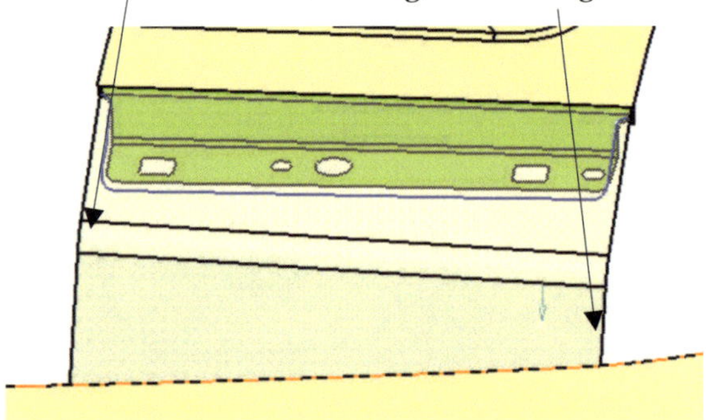

Fig. 22. Addendum of fender B that reused parameters of fender A

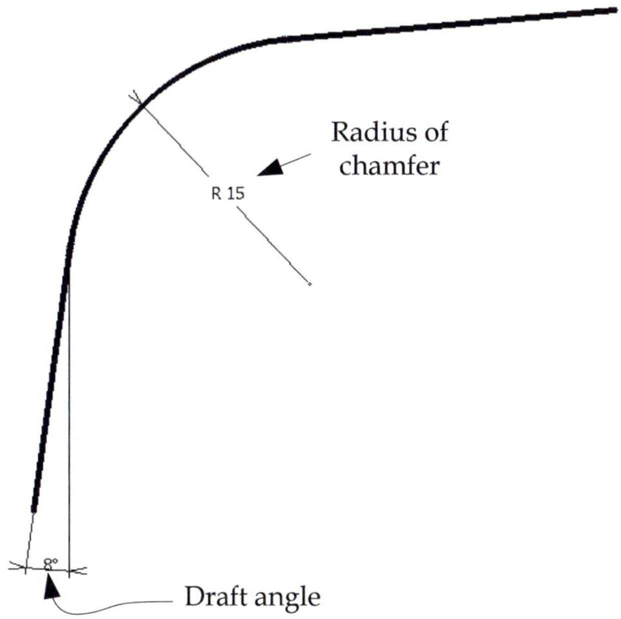

Fig. 23. The features that reused parameters of previous case

Take hood B as a global reuse example. Fig. 24-Fig. 27 show hood A and hood B and their section views. Considerable similarity exists between hood B and hood A; thus, many parameters of hood A are adopted for hood B such as type of the section curve, radius of chamfer, draft angle, addendum design, and trimming angle (Fig. 28).

Fig. 24. Hood A

Fig. 25. Hood B

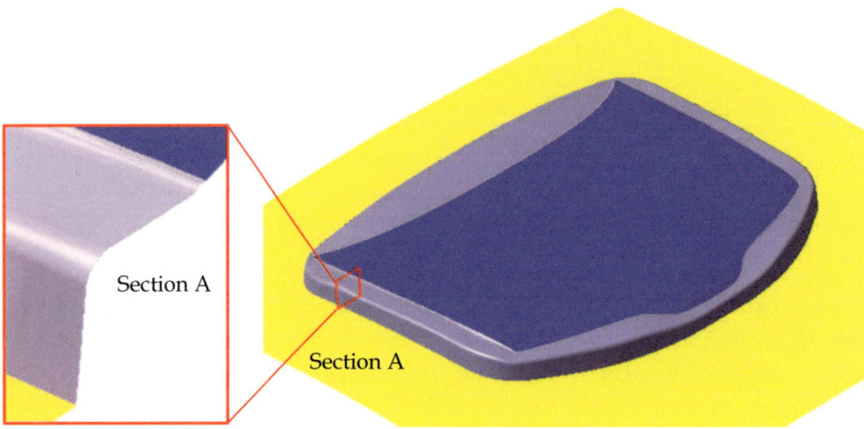

Fig. 26. Section view of addendum of hood A

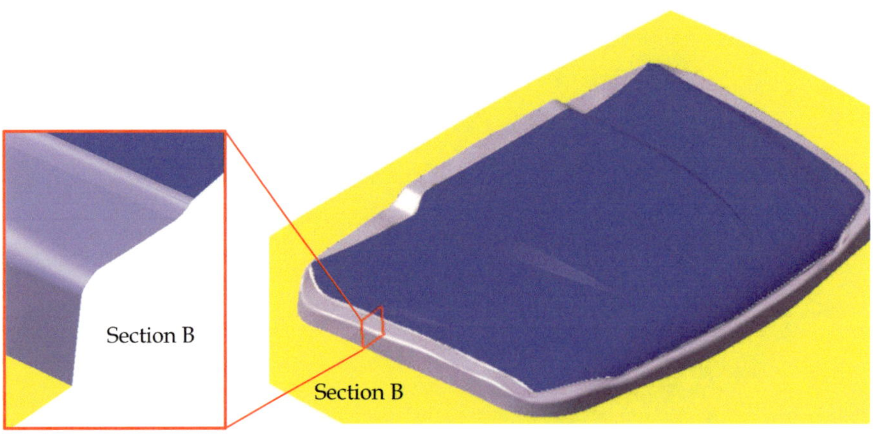

Fig. 27. Section view of addendum of hood B that reused parameters of hood A

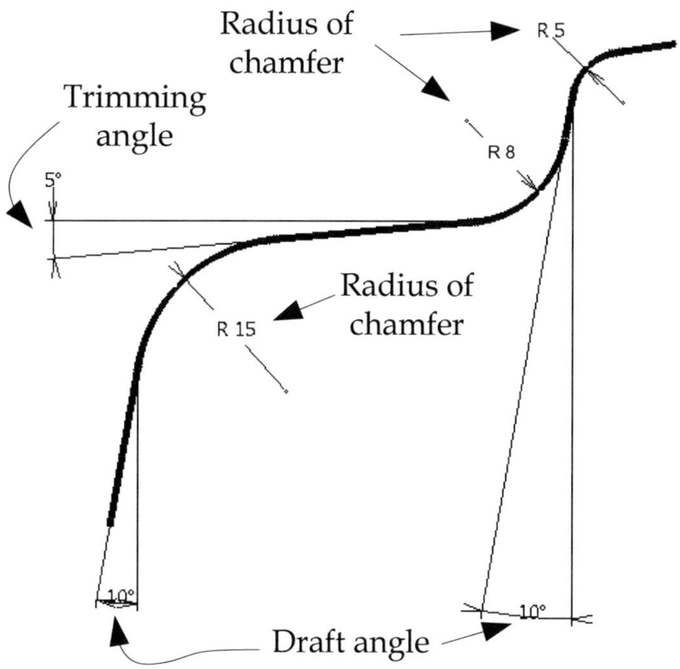

Fig. 28. The features that reused parameters of previous case

4. Conclusion and discussion

This study developed a framework in which KBE replaces the conventional method of designing a die face that relies heavily on designer experience and repeated trial and error, especially for addendum design due to the unpredictability of metal blank flow. With KBE techniques, designers can apply parameters from similar cases to a new case. In this study, feature recognition and case representation use features to describe cases and CBR compares features to determine their similarity.

The questions faced after integrating KBE into the design process of a die face is that, due to the geometric complexity of a die face, reuse may not be successful. Additionally, reuse relies heavily on retrieved cases from a case database, meaning that if previous cases are poor or the case database is biased, reuse is not desirable.

In the future, as the number of cases in the case database increases, CBR can search among an increasing number of possibilities to find the most similar case, resulting in desirable reuse. Furthermore, the reuse concept can be extended to the entire die design process, such that convenience for die designers can be increased.

5. Acknowledgement

The authors would like to thank the engineers of Gordon Auto Body Parts Co., Ltd. for their technical supports. Ted Knoy is appreciated for his editorial assistance.The authors would like to thank the National Science Council of the Republic of China, Taiwan, for financially supporting this research under Contract No. NSC-100-2221-E-002-061-MY2.

6. References

Dy, C.; Liu R.; Hu, P. & Song Y. (2008). Smoothing parametric method to design addendum surface, *International Conference on Intelligent Computation Technology and Automation*

Kendal, S.L. & Creen, M. (2007). *An introduction to knowledge engineering*, Springer-Verlag London Limited, ISBN 1846284759

Makinouchi, A. (1996). Sheet metal forming simulation in industry, *Journal of Materials Processing Technology*, Vol.60, pp. 19–26, ISSN 0924-0136

Polanyi, M. (1958) *Personal knowledge: toward a post-critical philosophy*, The University of Chicago Press, Chicago

Schenk, O. and Hillmann, M. (2004). Optimal design of metal forming die surfaces with evolution strategies, *Journal Computers & Structures*, Vol.82, Issues20-21, pp. 1695-1705, ISSN 0045-7949

Tor, S.B.; Britton, G.A. & Zhang, W.Y. (2003). Indexing and retrieval in metal stamping die design using case-based reasoning, *Journal of Computing and Information Science in Engineering*, Vol.3, Issue 4, pp. 353-362

Watson, I. & Marir, F. (1994). Case-based reasoning: A review, *Knowledge Engineering Review*, Vol.9, No.4, pp. 1-34

Zheng, J.; Wang, Y. & Li, Z. (2007). KBE-based stamping process paths generated for automobile panels, *International Journal of Advanced Manufacturing Technology* Vol.31 No.7, pp. 663–672, ISSN 0268-3768

3

Product Instructions in the Digital Age

Dian Li, Tom Cassidy and David Bromilow
University of Leeds
United Kingdom

1. Introduction

Product instructions are guides with the purpose of helping consumers to use products properly when they cannot be communicated through the design of the product itself. They are usually "on the product itself or its packaging or in accompanying materials" (ISO/IEC GUIDE 37, 1995, iv). They perform many different tasks and the good ones benefit both the users and the manufacturers. They should ensure users can operate products properly and safely. For manufacturers, instructions can add value to the products, encourage sales and reduce time for customer service which makes good business sense.

In recent times, products tend to be designed for intuitive use. However, this is not the case for every user and for every product. Thus product instructions are still necessary and have their unique values. For example, many newly developed products are very complex and involve multiple functions, these functions are experimented with only when necessary and when the user guides are available. Thus, instructions for these products have to be carefully prepared if the current trends are to continue.

In this research, the authors investigate problems of product instructions as well as suggest possible explanations for those problems. The study aims to find solutions to enhance the effective and inclusive performance of product instructions in a commercial environment in this digital age. The ideal was to find a method to produce durable product instructions that can be easily accessed, understood, stored and updated for all; meanwhile fulfilling the requirements of being cheap to produce and environmentally friendly.

2. Product instructions

In the Oxford English online dictionary (2006), the word "product" is defined as "that which is produced by any action, operation, or work; a production; the result. Now that which is produced commercially for sale". Another word "Instruction" is described as "making known to a person what he is required to do; a direction, an order, a mandate (oral or written) (www.oed.com, 2006)". Thus the term "Product instructions" refers to the guides associated with products to provide detailed operating instructions. In one of these international standards, "Instructions for the use of products of consumer interest" are defined as: "the means of conveying information to the user on how to use the product in a correct and safe manner" (ISO/IEC GUIDE 37, 1995, iv).

The initial purpose of instructions is to communicate vital information to users, and help them to use products correctly when this cannot be achieved through the design of products themselves. They are crucial parts of products and they should allow and promote proper use of manufactured goods also offer direct help to avoid mishandling which may lead to danger. Although they should not compensate for flaws of product design, instructions should be able to reduce risks of damaging products, consequent failures or inefficient operations.

It is believed by Petterson (2002) that the design of product instructions is the design of instructional messages and it is one sub area of Instruction Design. It is closely related to information design and it is an interdisciplinary subject. It takes influences from many established areas of research. The main areas may involve language, art and aesthetics, information discipline, communication, behavioural and cognitive study and so on.

In many perspectives, product instruction design and information design share their similarities. Both these areas are not clearly defined yet and their histories are not easy to trace. The preparation, presentation, analysis and understanding of a message need to be embraced through a selected medium. Also, product instruction design and information design both involve multi-disciplinary and global concerns. They have influences from similar areas and they both need to inform the intended users. When studying the design of product instructions, the authors have taken inspiration from the information design field.

3. The trends and problems

People's demands on product instructions more or less depend on how much information they need for the operation of the products. Therefore the design of product instructions is closely related to the design and development of products. In the product design field, products are restructured and redesigned from time to time to follow different trends. Designers are passionate about using newer technology, creating appealing appearances for products, minimise then simplify them (Redhead, 2000).

The past ten years has been the fastest changing period for technologies; extraordinary changes have been brought to society. The emergence of mass information, new products and redesigned products is more prevalent than ever. New technology allows products to be more complicated to use but their appearance is getting simpler. Users are very often overwhelmed by new technology, products and information. Products both more confusing and more exciting than ever.

The choices for customers are wide,from products like computers and mobile phones to consumer products, for example, shampoo and chairs, almost everything has been rethought from scratch. Consumers, after experiencing high technologies and pleasant appearances of inventions, start to require more control and urge for personalisation. They want products to be "more seductive, more personal, more available" (Redhead, 2000, P54). The trend of customisation allows more customers to personalise their products. For example, Apple's products such as the iPod and iShuffle allow customers to have their own name on the back of the product. Similarly, Nike let the buyers decide the colour, detail and the ID of their own trainers. And now, personalised products are developed in most industries; ranging from fashion products to computers or furniture, even picture frames.

These trends and existing changes on the design of new products require product instructions to follow up and to be as exciting. The change of products needs their instructions to be clearer, quicker to access and easier to understand. However, while we are surfing in the ocean of newer and more charming products, product instructions are not as interesting as they should be. They are not redesigned and updated quickly enough to catch up with the changes in emerging products. With the majority of product instructions, users are mainly suffering from problems in three design aspects: the effectiveness, the accessibility and the inclusiveness.

3.1 Effectiveness

Although product instructions are important and not replaceable, evidence suggests that existing instructions are not as good as they should be. The fact is that the user satisfaction rate on product instructions is only 31% according to the authors' survey in 2006. Users complain about many problems with product instructions such as they do not explain what the users really need; they are either too wordy or difficult to understand, hold unusual technical terms, contain bad translations and bad visuals (Figure 1).

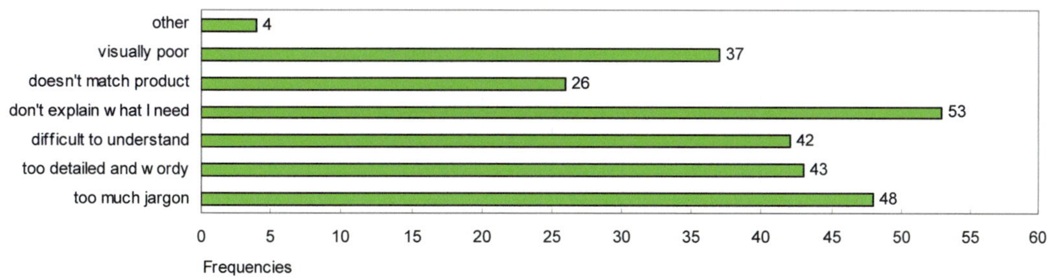

Fig. 1. Problems with instructions (authors' survey 2006).

These criticisms suggest that many accompanying instructions are not effective and they are not designed for all.

3.2 Accessibility

The majority of instructions in accompanying materials are in the physical forms of leaflets, manuals, CD (Figure 2). Another survey carried out by the authors (2008) showed that the vast majority (92%) of product instructions often are printed. Other forms such as CD/DVDs are also adopted but only used by a very small number of users.

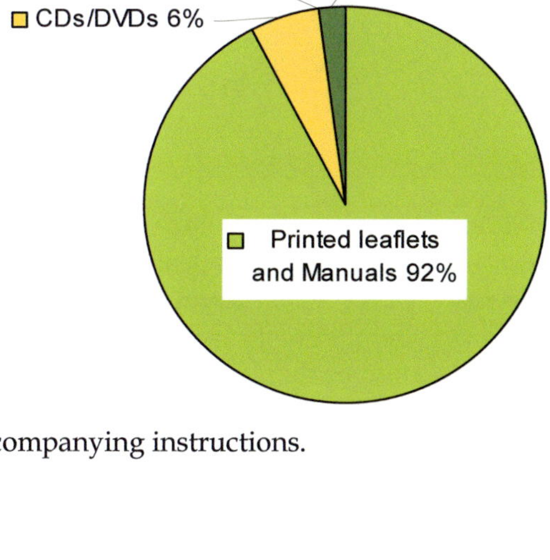

Fig. 2. Types of accompanying instructions.

Many printed instructions take a huge amount of storage space and are not easy to be kept and shared. For example, some products, such as office machines and equipment, are shared or passed around and instruction manuals become lost in the process. For example, in the survey carried out by the authors (2008), 72% of participants (155) intended to keep all instructions which accompany products (Figure 3).

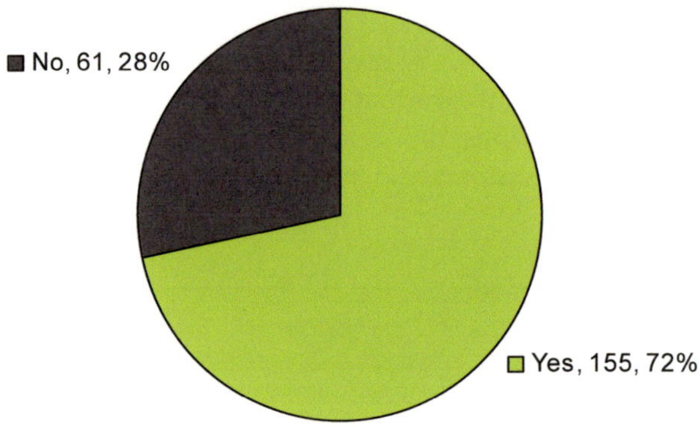

Fig. 3. Do users intend to keep instructions?

Among them, only 5% participants (3%) never lose the instructions and another 37 participants (24%) rarely lose them. 12% participants admit that they lose product instructions very often. The majority users (61%) replied "sometimes" (Figure 4).

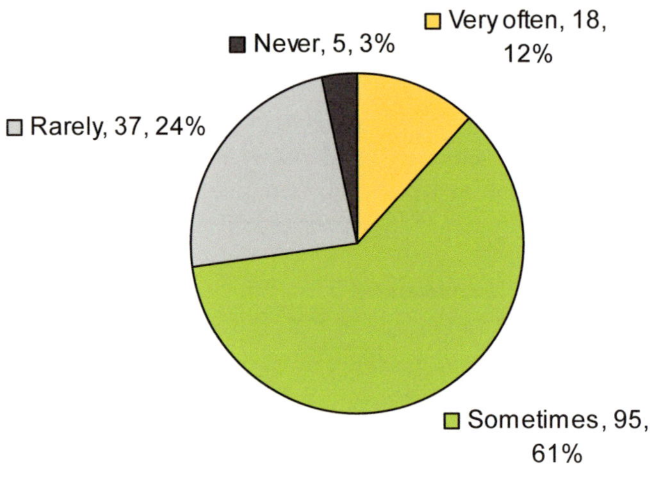

Fig. 4. Do users lose instructions?

On the other hand, for manufacturers, it is expensive to have these accompanying materials produced. Paper based product instructions such as leaflets and manuals are typical examples. Some of them contain many pages as all useful information has to be included (Figure 5).

Costs for producing them have been continuously increasing since the price of energy and paper has climbed. In recent years, there are more and more product instructions available online for free download. However, they are mostly digital or scanned versions of the traditional instructions, and are still limited compared to other instructions delivered by physical media.

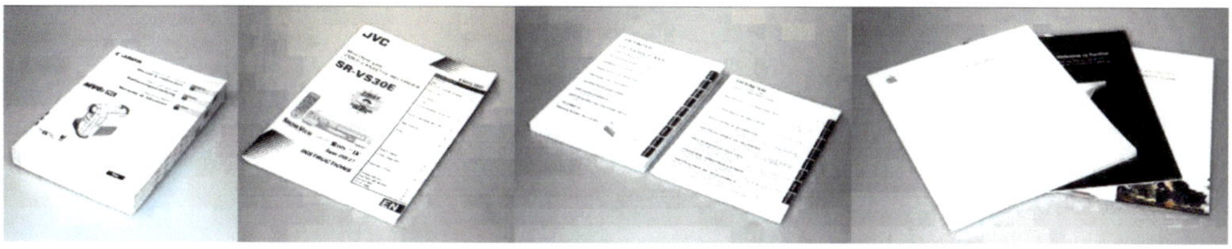

Fig. 5. Examples of paper based product instructions.

3.3 Inclusiveness

Product instructions should help all users to use different products and fulfill different tasks. When designing product instructions, designers are planning instructional materials and in some way designing a learning process for product users. Designed instructions should be easy to follow by users with all kind of intelligence references and learning styles. Therefore the instructions should use varied styles of delivery, be customized for specific people and help everyone to understand and learn. However, this is not achieved in most cases.

Currently, the majority of product instructions are presented by either text, images or a mixture of both. Product instructions dominated by text certainly enable aural learners to follow easily, however, they are not as straightforward to use for other readers with different strengths. Instructions full of pictures and charts can help visual learners to process information effectively and use products quickly but yet again they might not be the best choice for other users, for example, kinesthetic learners.

Additionally, it is difficult to locate a piece of information among instructions, as they are often very lengthy. Users have to scan all information and evaluate them to select the part they need. This works well when the assimilating learning style is favoured. However, when an accommodating learning style is preferred, users will be annoyed, often because they are forced to go through lots of instructions, instead of getting hands on experience quickly.

4. Current solutions

Facing all these problems with product instructions, some actions have been taken to alleviate the frustrations. To make product instructions more comprehensible and effective, standards for formulating instructions are available and textual materials on how to write instructions are provided. Meanwhile, info-graphics have been studied by some designers and academic researchers so that graphics can be used to aid the presentation of information. However, related standards are limited and dated; research focused on the accessibility and inclusive design of product instruction is very rare; problems of product instructions are not completely and successfully solved and users are continuing to suffer from annoyance caused by poor product instructions. It is necessary to carry out a systematic and up to date study to improve the performance of product instructions, especially in this digital age.

5. Possible solutions

The people who are producing product instructions should be trained to write and produce effective product instructions. They should understand standards for formulating instructions and be able to apply them while doing their jobs.

To make product instructions easily accessible, easily stored and updated, they could be created and distributed digitally, through networks, for example Internet or 3G networks. This will provide product instructions available at anytime, from anywhere around the world and could be translated into multiple languages with a very low budget for maintenance. There were 1,966,514,816 Internet users around the world in June 2010. The number has grown by 444.8 % between 2000 to 2010, and it is still growing (internetworldstats.com, 2010). On the other hand, based on Nielsen's estimate (2010), 50% of US mobile subscribers (142.8 million) will be Smartphone users by 2011, which means they could get access to a 3G network very easily on their phones. Similar trends are actually happening everywhere across the world.

To fulfil requirements from users with different intelligence levels and learning styles, instructions could involve multiple media for example, sound, music, animation etc., as well as the traditional media of text and images. Product instructions might also be interactive so that they could be read in almost any order. Once instructions are designed to be interactive rather than linear, they can be read by choice. This should enable the users to reread instructions and to repeat the tasks when an error is discovered. This will also minimise the amount of time spent on reading instructions, especially for those inexperienced users who have little prior knowledge. Also, a combination of minimalist and systematically complete instructions might be able to offer the most productive learning experience.

6. Design challenges

Ideally, multimedia instructions should help people with different leaning styles and strengths to operate products easier, quicker and safer. However, it has been suggested by Tapscott (2009) that Digital Natives (those who have grown up with digital devices) and Digital Immigrants (those who learnt to use digital devices as an adult) learn things differently and have different opinions on digital products and interactive works. Therefore the key challenges for this study were to find differences between improved traditional instructions and multimedia instructions in use; to discover if multimedia instructions are going to perform better in terms of their effectiveness and inclusiveness; to determine if multimedia instructions can be better solutions for all users, including the Digital Natives and the Digital Immigrants.

7. Prototyping and user testing

To find the answers to the above questions, instructions for a particular product, a lighting table (Figure 6) were chosen and rewritten according to the standards and regulations for planning product instructions. Two versions of product instructions were then produced: a printed version combining text and images (Figure 7); a multimedia version of the product instructions, which used the same text and imagery information but involved extra sound, animation and were designed as interactive (Figure 8). Participants for the tests were separated into two groups: Digital Natives and Digital Immigrants.

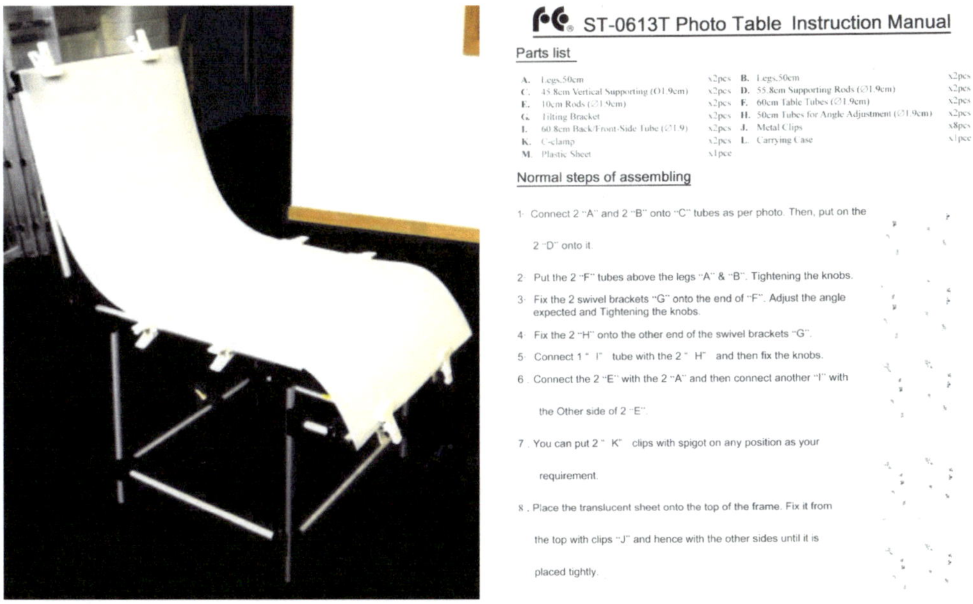

Fig. 6. A photo of the selected product and its original instructions.

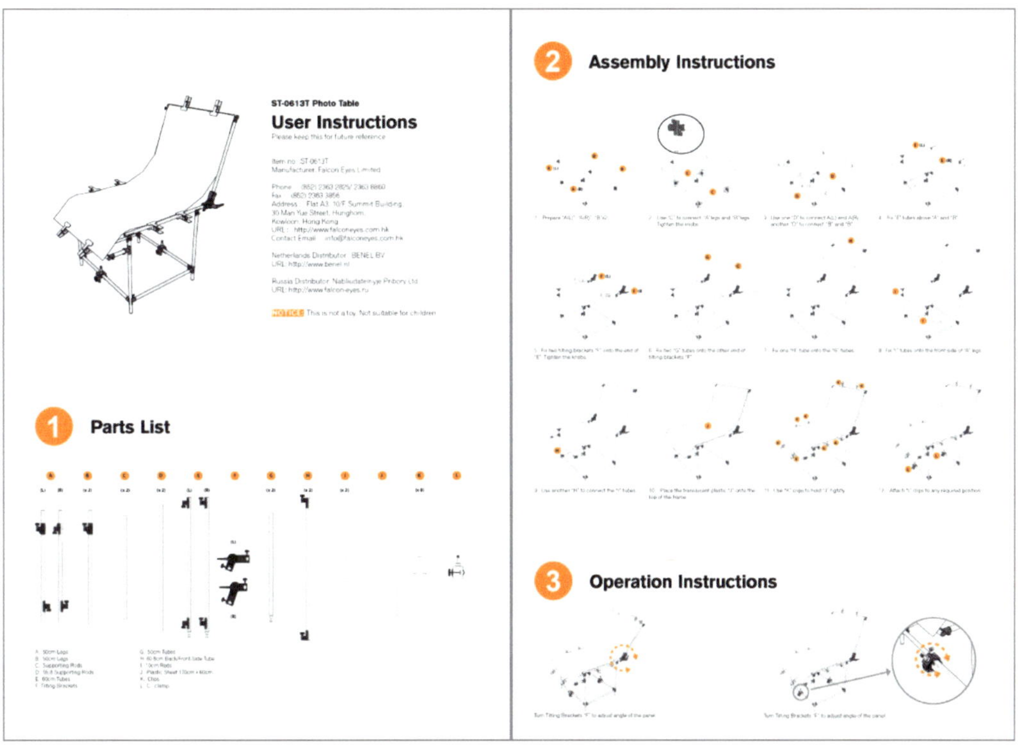

Fig. 7. Layout for the redesigned printed version of product instructions.

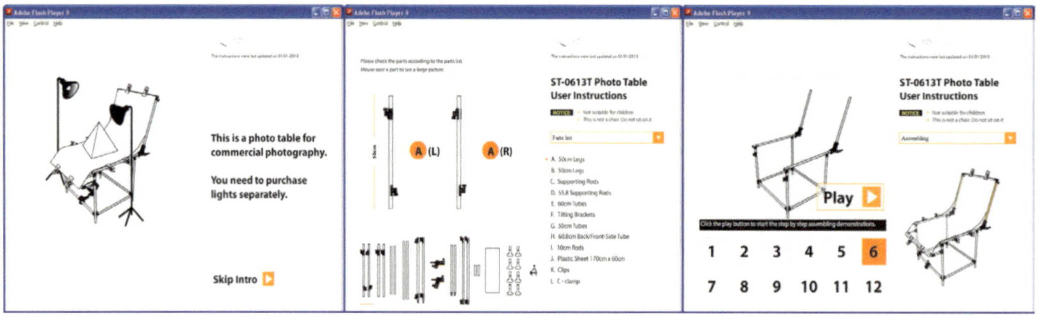

Fig. 8. Screen examples for the multimedia version of product instructions.

The choices of the participants had to be carefully planned in order to ensure that their human performance characteristics were checked prior to the beginning of the tests. This was done using a method traditionally used by work-study officers to establish what a hundred percent effort or rating looks like. A full pack of cards is dealt into four hands in a period of 52 seconds. It was ensured that the participants could all carry out this task in periods between 50 and 56 seconds. Thus their human performance characteristics could all be considered similar.

They were asked to follow either the printed instructions for the product or the multimedia ones to complete the same set of given tasks. The tasks included 1) using instructions to find information, 2) checking components of the product and 3) assembling. Their actions were monitored; timing was recorded; errors were observed and feedback was collected at the end.

8. Analysis and interpretation

Data from the experiment was gathered and interpreted to clarify the communicating effects of instructions, mainly from two aspects of task performance: efficiency and accuracy. Results from different user groups were compared and examined to find out which type of instructions were more effective and inclusive in terms of communication.

8.1 Task analysis

In the test, all components of the product were given to the participant and product instructions were provided on either a printed sheet or a laptop computer. By using Hierarchical task analysis (HTA), the main goal and sub-goals of the tasks were expressed as below (Figure 9).

As shown in figure9, tasks 1 and 2 were skill-based tasks. Users were asked to follow clear and simple instructions to carry out actions and they did not have to make judgements on the aim and plan of their actions. Task 3 was more complicated, it contained 12 steps and these 12 little tasks were grouped into different sections: 3-A, 3-B, 3-C and 3-D, depending on their goals. Many parts of task 3 required users to recall or understand some rules then make their own decisions on what action they should carry out, in order to achieve the required results.

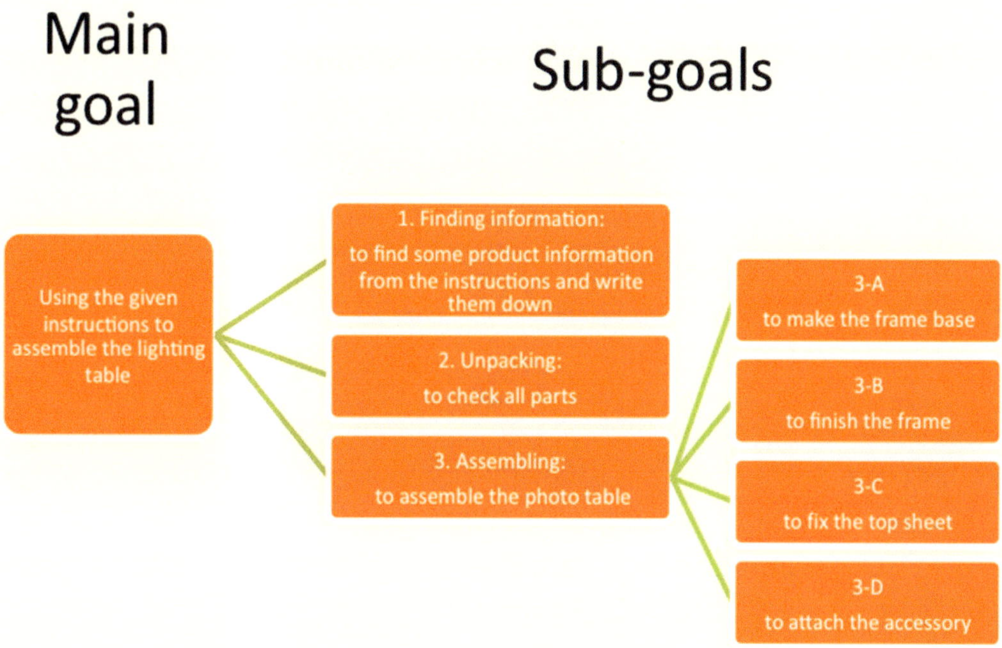

Fig. 9. The main goal and sub-goals of the refined tasks.

8.2 Performance on efficiency

To examine efficiency of different types of instructions, the following indices were used:

1. Time for finishing all tasks, (from the beginning to the end).
2. Time for each task section.

Overall, Data suggested that all participants, who used multimedia instructions, used on average 1379 seconds (22minutes 59 seconds), with a standard deviation of 220 seconds, which seemed to be slightly shorter to those of printed instruction users, who spent 1390 seconds (23 minutes 10 seconds), with a standard deviation of 213 seconds (Table 1). The difference was small therefore the authors were keen to find out if there was statistically significant difference between the total time consumption of different instruction.

Null hypothesis: $T_1 = T_2$

There is no significant difference between the average total time of users using either printed or multimedia instructions.

Alternative hypothesis: $T_1 \neq T_2$

There is a significant difference between the average total time of users using either printed or multimedia instructions.

The authors assumed that there was no statistically significant difference between the groups. A t-test was then carried out. to find out the confidence level of this hypothesis. The 2-tailed t- test returned a P-value of 0.89. Since the P-value (0.89) was greater than the significance level (0.05), the null hypothesis was accepted. This indicated that there was no significant difference between the efficiency of two versions of product instructions.

		Digital Natives (DN)	Digital Immigrants (DI)
		Average Duration (s)	Average Duration (s)
Printed product instructions	Average duration (s): ≅1390	Printed (DN) ≅1284	Printed(DI) ≅1497
Multimedia product instructions	Average duration (s): ≅1375	Multimedia (DN) ≅1332	Multimedia (DI) ≅1425
		All DN Average(s) ≅1308	All DI Average(s) ≅1461

Table 1. Average time for all tasks

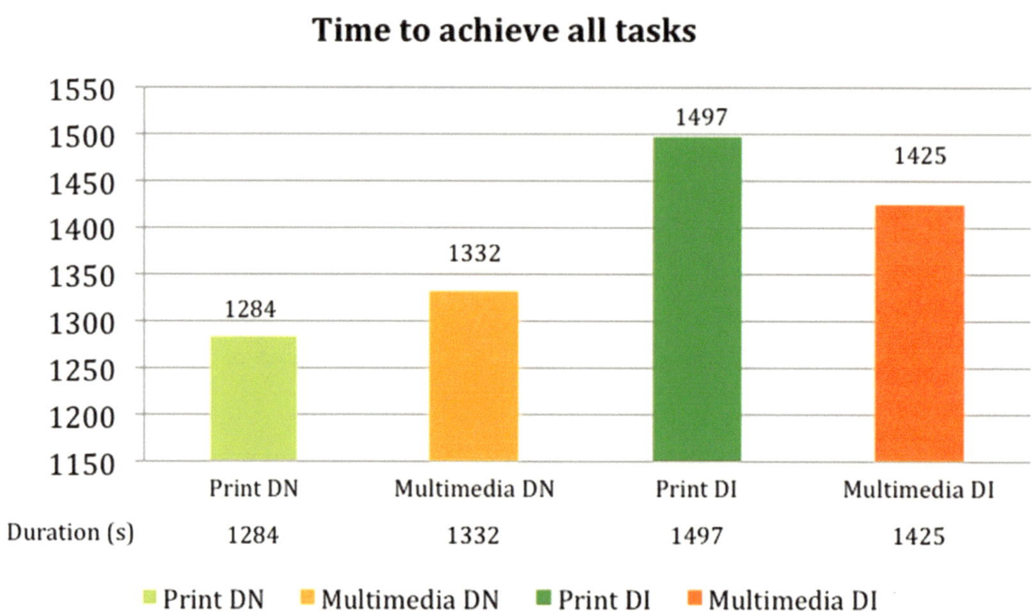

Fig. 10. Time to achieve all tasks in each user group.

Similarly, Figure 10 potentially suggested that the Digital Natives performed better, when using either the printed or multimedia instructions, than the digital immigrants. On the other hand, the Digital Natives seemed to perform better using print instructions than multimedia and the opposite happened with the Digital Immigrants. Again, t- tests were carried out to find out the confidence level of the significant differences.
After another 3 sets of 2-tailed t-tests, the results showed that the amount of time Digital Natives and Digital Immigrants used to complete all tasks was not much different. Also, different types of instructions did not have a huge impact in terms of the overall efficiency. This suggested that multimedia product instructions were as efficient as printed ones for both Digital Natives and Digital Immigrants.

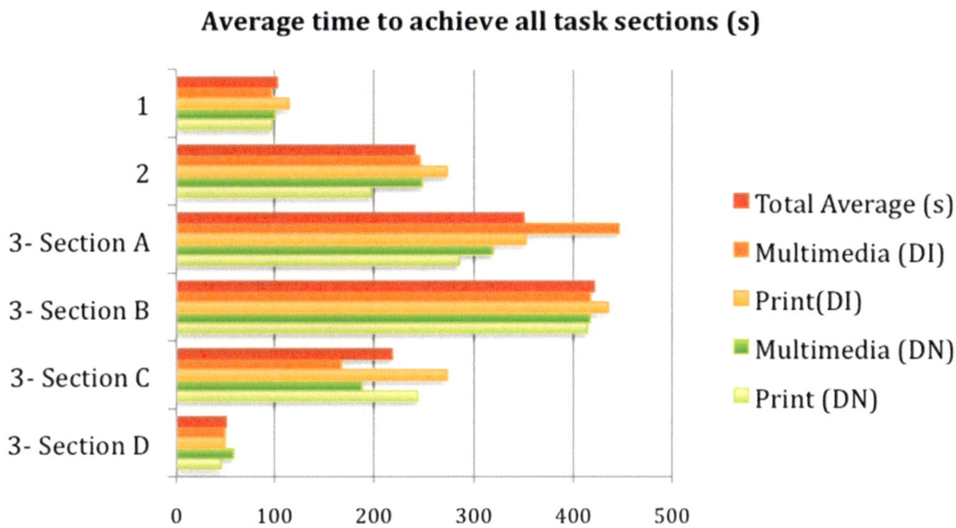

Fig. 11. Average time to achieve all task sections in each user group.

By looking at the time, which has been spent on each task section in detail (Figure 11), it was discovered that multimedia product instructions generally have a positive impact on Digital Immigrants in terms of their efficiency. Yet, this does not apply when a spatial judgement is involved in knowledge- based tasks. For Digital Natives, different types of instructions have different effects on time consumption; depending on the nature of each task. Printed instructions were more efficient to be scanned when skill-based tasks were simple enough to be completed intuitively. When instructions are necessary to follow for skill- based tasks, the time difference between the use of multimedia instructions and printed instructions was very little. Although it also took longer for the Digital Natives to use multimedia instructions to fulfil a complicated job when a spatial judgement was needed; multimedia instructions were significantly more efficient to use for a transferable process (rule-based tasks). Therefore the authors believe that multimedia instructions are not always efficient as expected, for example for particular types of tasks; but they are still more effective in terms of their efficiency in general, especially for Digital Immigrants.

8.3 Performance on accuracy

The accuracy of users' performance was evaluated through error analysis in this investigation and it was achieved by observations. Every time a user failed to achieve the planned goal, a

human error was found and recorded. The failures mainly appeared in two different ways (Hollnagel, 1993): the user's plan was adequate but the actions were deficient; or actions were carried out as planned but the intention was wrong according to the given tasks.
To define and classify human error in the tasks, three questions were referenced (Reason, 1990), and they all closely related to intentions:

1. Were actions guided by prior goals?
2. Have actions been proceed as planned?
3. Was the desired end achieved?

The questions then were used to categorize discovered errors into three main kinds: slips, lapses and mistakes (Figure 12).

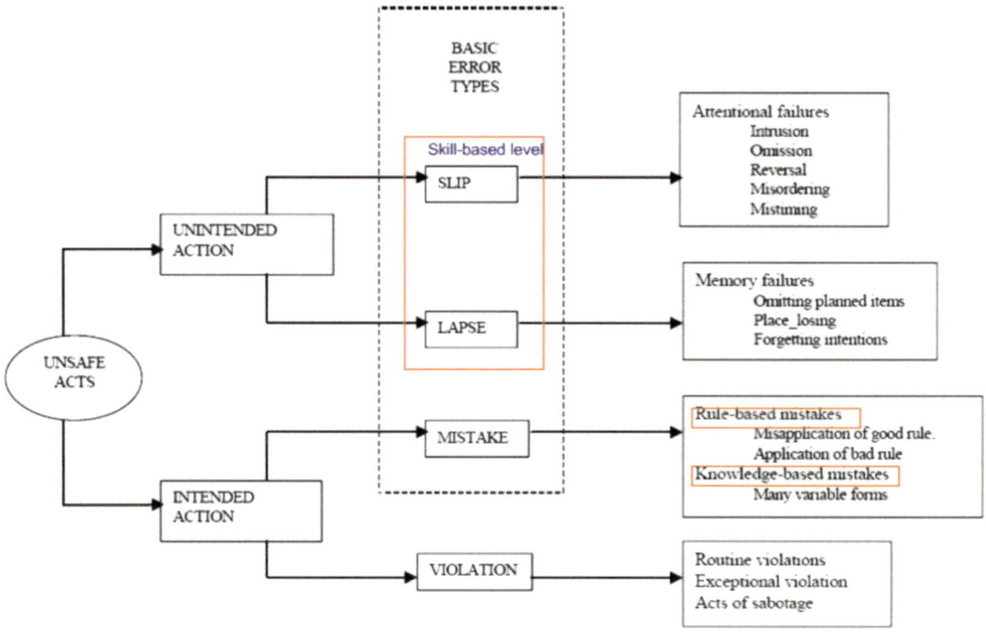

Fig. 12. Generic Error Modelling System (GEMS) (Reason, 1990)

8.3.1 Slips and lapses

Both slips and lapses were errors that happened in the execution process so that users failed to achieve the desired goals. The intentions were proper, but results were not right. A slip mainly involved actions that were not happening as planned, while lapses were the errors that were caused by failures of memory.

8.3.2 Mistakes

Mistakes referred to failures in the planning and/or judgemental processes; they are errors in users' intentions (Norman, 1983). They might involve the selection of objectives, or the decisions of means, even actions for achieving the objectives. They are normally caused by either a failure of expertise or lack of expertise(Reason, 1990). They could be effectively avoided or reduced by communicating sufficient information, in this case, using product instructions. Therefore, the number of mistakes left by users after each test was the main criteria to evaluate the accuracy performance of the product instructions.

	Print DN	Print DI	Multimedia DN	Multimedia DI
Slips	2	1	1	2
Lapses	0	1	0	0
Mistakes	4	2	1	0

Table 2. Error analysis

During the tests, by referring to given instructions, users observed and corrected their own errors in the performing progress. When errors were not detected or corrected by the participants, they were recorded and evaluated (Table 2). In this experiment, Digital Natives and Digital Immigrants have both made errors. The numbers of slips and lapses were similar in two user groups. They were caused by each individual's failures of action or failures of memory. Unfortunately, they cannot be perfectly solved by improving product instructions. When looking at the number of mistakes made by the participants, data revealed that multimedia instructions are better for avoiding mistakes, for both Digital Natives and Digital Immigrants compare to the traditional instructions.

	Tasks	Multimedia (DN)	Print (DN)	Multimedia (DI)	Print (DI)
1	Finding information	0	0	0	0
2	Unpacking- check parts	0	0	0	0
Section 3-A:	Make the frame base	1	0	0	0
Section 3-B:	Finish the frame	0	3	0	2
Section 3-C:	Fix the top sheet	0	1	0	0
Section 3-D:	Attach the accessory	0	0	0	0

Table 3. Mistakes in each task section.

When each task section was examined (Table 3), it was more evident that multimedia instructions can help users to avoid mistakes, especially in some complicated tasks for example task section 3-B. This section contained the most time consuming jobs in the whole experiment. It was a combination of many small knowledge-based tasks and some tasks required users to make critical decisions. Both Digital Natives and Digital Immigrants have made a few mistakes in this job section when traditional instructions were provided. However, when multimedia instructions with equivalent contents were presented, all users either avoided mistakes or realised errors on their own. Thus it is not difficult to conclude that multimedia instructions are better than traditional instructions (with only images and text) in terms of ensuring accuracy of actions.

9. Conclusion

Although multimedia instructions are not always more efficient than traditional ones, the authors believe that they are quicker to use for many types of tasks. They allow more flexibility in action and more importantly can help to reduce human errors significantly in the working process. They are effortless to access, can be transferred between different digital platforms and easy to update or replace. It is also proved that they can be used easily by people who have grown up with or without ubiquitous of digital media.

During the time for the development of this project, technology has changed and improved as fast as usual. Smart phones have become more popular and more portable digital devices like the iPad have come into the market and our daily lives. We are more comfortable with and dependent on with portable digital devices. The authors believe that multimedia instructions could be and soon will be better solutions for many of our daily products, to help us use products easily and safely.

However, this will not cause the complete death of the traditional instructions. Although they are not as good to ensure accuracy of performance, they can still be used in many cases. For example, they are a better choice for performing simple skill-based tasks on products in terms of efficiency, especially for Digital Natives, less patient or busy people. Therefore, in the future, instruction designers may want to analyse tasks for using a particular product, then decide either they need to design traditional instructions or multimedia ones to satisfy users' requirements for efficiency and accuracy of operation.

The authors accept that, as with any research project, there are certain limitations associated with the study. Every effort was made to ensure that the participants chosen had similar levels of human performance characteristics. In other words, they all had the same degree of normal flexibility, intelligence, adaptability and dexterity. However, even with the test applied (dealing cards) and observations made on the participants, it was still impossible to completely eliminate the human factor.

10. Contribution

Overall, for this study, a problem-driven design strategy was adopted. Inspiration was taken from fields like information design, product design, graphic design, and instructional design in education plus cognitive and ergonomic science.

As related studies and standards are not sufficient to solve problems with product instructions especially in this digital age, this research should make an original contribution in this field. It will benefit both users and manufacturers since it aims at finding solutions to improve the quality of product instructions. Furthermore, this study should have economic value as the costs of product instructions can be reduced.

11. References

Entner, R. (2010). *Smartphones to Overtake Feature Phones in U.S. by 2011.* October 27, 2010, Available from: <http://blog.nielsen.com/nielsenwire/consumer/smartphones-to-overtake-feature-phones-in-u-s-by-2011>

Hollnagel, E. (1993) *Handbook of cognitive task design,* Mahwah, N.J. ; London : Lawrence Erlbaum.

ISO (1995). ISO/IEC GUIDE 37: 1995(E) Instruction *for use of products of consumer interest.*

Miniwatts Marketing Group. (2010). *Internet usage statistics - The Internet Big Picture.* October 25 2010, Available from: <http://www.internetworldstats.com/stats.htm>

Norman, A.D. (1983). Design Rules Based on Analyses of Human Error. *Communication of the ACM,* Vol 26 254-258

OED. (2006). *Oxford English online dictionary* , January 14 2006, Available from: <www.oed.com>

Pettersson, R. (2002.) *Information design: an introduction,* Amsterdam, Philadelphia : John Benjamins Pub. Co., c.

Reason, J. T. (1990). *Human error* , New York: Cambridge University Press

Redhead, D. (2000) *Products of our time,* London, Birlhauser.

Tapscott, D. (2009) Grown up digital: how the net generation is changing your world, New York ; London : McGraw-Hill.

Part 2

Culturally Inspired Design

4

Culturally Inspired Design: Product Personalities to Capture Cultural Aspects

Denis A. Coelho, Ana S. C. Silva and Carla S. M. Simão
Universidade da Beira Interior
Portugal

1. Introduction

This chapter, focusing on culturally inspired design, with emphasis on Portuguese and Lusophone cultures, is developed in a two stage process (Fig. 1). In the first part, an effort to identify the Portuguese identity reflected in the design of existing products is pursued. In the second part of this work, product design specifications are created based on the assignment of product personalities to capture Portuguese and Lusophone cultural aspects. Both stages of this contribution give rise to new product concepts, which are aimed at exemplifying the profile in existing Lusophone design production (in comparison with other design origins) and at demonstrating the transfer of selected cultural values to designed objects.

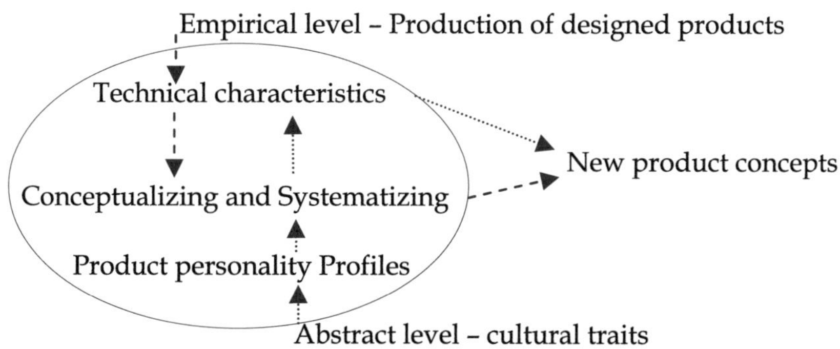

Fig. 1. Depiction of the two streams of analysis departing form an empirical and an abstract level, reaching at new product concepts.

In both stages of the research, an array of product features was drawn up, in the first case from observation, in the second case from matching of cultural traits with product features, through the use of the product personality assignment approach.
Culture may inform design by a process of context-informed practice. Hence, collectively-held norms of practice shared within contexts may well introduce culture into the design process, even if indirectly. Geographical context may influence the practice and results of design in two ways. One the one hand, everyday specific features of a location (availability of technology and materials, climate, local modes of exchange and even cultural factors affecting business activities) produce particularized actions, which may however be

contrasted with perceived globalized, mainstream and dominant modes of practice. On the other hand, when viewing design as a mode of communication, peripherization and engagement of a consciousness of difference may emerge, depending on location (Julier 2007). No factual data with a substantial depth about the cultural traits of the nations portrayed in this chapter was found in literature, with the exception of the work by the Dutch social scientist Geert Hofstede. Hofstede developed and published, in 1980, four national measures of culture applied to a set of selected countries. Portugal and Brazil were the only Lusophone countries included in Hofstede's study. The national measures of culture presented by this author were: Power Distance, Masculinity, Individuality and Uncertainly Avoidance. The nature of the national measures of culture presented by Geert Hendrik Hofstede, was not deemed adequate to advance the development of the goals set for the second project reported in this chapter. A literature survey was hence pursued, informing, through the perspectives of several authors, on the Portuguese and Lusophone cultural traits.

The current geo-strategic setting gives some added importance to the Portuguese-speaking world. Both Brazil and Angola, in part, have been claiming themselves as regional powers (in South America and sub-Saharan Africa, respectively). In this context, the design of products as part of the cultural expression of people is associated with its production and industrial capacity, and can be seen as a front for disseminating advancement of culture, while its existence is related to the relative importance of this culture in the globalized world. It is not mandatory to join Brazilian design, and eventually Angolan, or the design of other Portuguese-speaking countries, with Portuguese design. In the cultural sphere, and the approach that underlies this work has a strong cultural component, it will be difficult to disentangle the historical and cultural legacy of the area of language, as this is one of the main ways to define and mark broad cultural groups. Thus, in this work, it is considered that the combination of design production in the countries of official Portuguese language is relevant.

1.1 Aims

The main purpose of the first part of this chapter is to seek the identification of a possibly existent identity of Portuguese and Lusophone Design, according to different perspectives (e.g. form, brand, material, archetype), from the study of selected cases. While it is acknowledged that an analysis which is mostly based on the material properties of products is necessarily limited in scope, the consideration of experience or use related qualities, given the breadth of this survey, was inferred, albeit visually as their assessment by actual use would not be feasible. Initially, a historical perspective of Portuguese and Brazilian design was drawn up. Since the existing information concerning existing design of other Lusophone countries is very limited, it was chosen to analyse Portuguese and Brazilian design only, and from these two identities, extrapolate a proposed Lusophone design identity, focusing on material properties mostly. From this analysis, similarities were perceived between Portuguese and Brazilian (Lusophone) designs, according to the analyzed products. Another analysis of designed products was then carried out, focusing on countries with design production of great international appreciation so that it would be possible to differentiate this against Lusophone design. The analyzed regions and countries, for the purpose of differentiation, were Scandinavia, which includes the Nordic countries, as well as Italy and Germany. To conclude the first stage of this research, and project it in a practical component, two conceptual designs were developed.

The second part of this chapter reports on a project that aimed to identify the cultural traits of the Portuguese speaking countries, with regard to both an internal perspective as well as an outsider's perspective. Subsequently, the translation of these traits into product design was intended, attempting to give a Portuguese and Lusophone projected cultural identity to products. To this end, a methodology was developed in several stages. For the application of the methodology, several studies were carried out. The personality attributes of products were analyzed using a technique known as Product Personality Assignment (Jordan 2000) in order to mediate the transfer from the identified cultural traits to product design requirements. Patrick W. Jordan used positive and negative characteristics of people, developing a list with 209 descriptors of personality and, after a collation and synthesis of work arrived at a list of 17 pairs of dimensions of personality. These dimensions are composed of pairs of opposing personality descriptors, such as Complex / Simple. Thus, Personality Assignment to a product is a tool that explores the emotional ties existing in the relationship between user and product.

2. Characterization of the identity of existing Portuguese and Brazilian design

In this section, an attempt to identify the Portuguese and Brazilian identities reflected in the design of existing products is carried out. This contribution gives rise to new concepts, which are aimed at representing cultural traits embedded in objects. An array of product features is drawn up from observation of a sample of designed objects (208), whose pictures were readily available from design web-blogs and design museums which were found through web searches, to empirically assess the existence of a Lusophone design style, in comparison with other design origins. The main purpose of this section is to seek the identification of a possibly existent identity of Portuguese and Lusophone Design, according to different perspectives (e.g. form, brand, material, archetype), from the study of selected cases. While it is acknowledged that an analysis which is mostly based on the material properties of products is necessarily limited in scope, the consideration of experience or use related qualities, given the breadth of this survey, was inferred, albeit visually as their assessment by actual use would not be feasible. Initially, a historical perspective of Portuguese and Brazilian design was drawn up. Since the existing information concerning existing designs of other Lusophone countries is very limited, it was chosen to analyse Portuguese and Brazilian design only, and from these two identities, extrapolate a proposed Lusophone design identity, focusing on material properties mostly. From this analysis, similarities were perceived between Portuguese and Brazilian (Lusophone) designs, according to the analyzed products. Another analysis of designed products was then carried out, focusing on countries with design production of great international appreciation so that it would be possible to differentiate this against Lusophone design. The analyzed regions and countries, for the purpose of differentiation, were Scandinavia, which includes the Nordic countries, as well as Italy and Germany. To conclude the first stage of this research, and apply it in a practical component, two conceptual designs were developed (Fig. 2 and 3).
One of the designs concerns a refrigerator (Figure 2) that intends to reflect the Portuguese identity, without disregard to new options, both in terms of currently available material and technology. The other conceptual design consists of a sofa with a special focus on Lusophone related features (Figure 3). The latter may assume an array of different configurations, and it differs from similar products in its versatility, having as main function that of a sofa, but including a footrest for the support of the users' feet, and converting into a set of table with three stools if necessary.

Fig. 2. Refrigerator designed with inspiration on the Portuguese "postigo" (small door or window within a regular door) (designed by the third author).

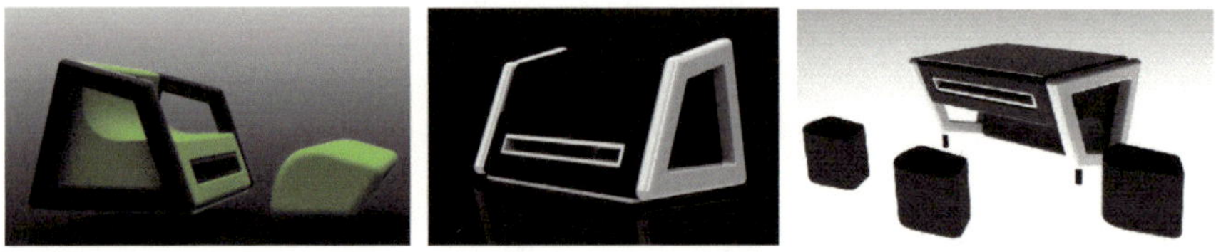

Fig. 3. Multi-purpose object (sofa, table, shelf, stools and footrest) designed with inspiration taken from the traditional "canapé" (multiple seat wooden chair) (designed by the third author).

2.1 Methods deployed to unveil existing design profiles

The overall goal of the study was to identify from various perspectives (brand, material, archetype, among others) the contours of a possibly existing identity of Lusophone design, from the study of selected cases. The guiding specific objectives were the following:

- Identifying the various types of associations that support cultural identity and seeking to illustrate them by adopting a historical perspective.
- Analyzing products of international recognition to identify a possible identity of Portuguese and Lusophone design.
- Placing the proposed identification of traces of Lusophone cultural identity in the context of other traditions, as a means of differentiation.
- Proposing solutions or concepts in continuity to what was found, while adopting contemporary or emerging technology.

To assist in achieving these objectives the following research questions were developed:

- Over time is there a continuity and perseverance in the appearance of traces, signs or marks on the production of objects within the Lusophone space, and in Portugal?
- Are there materials, shapes, graphic markings, colours, and other product properties typically Portuguese (Lusophone), and, or, with international acceptance?
- Are there any identifiable differences between the products of Lusophone production and the most visible design currents with a geographical identity, such as Scandinavian, Italian or German design?

An extensive review of Portuguese and Lusophone design was carried out in order to better understand it. The new designs created in the course of this study are based on results from the analysis pertaining to the products shown in the following sections. The selection of products comprised in the analysis presented (including iconic designs identified in design web-blogs, items for sale at the Museum of Modern Art, New York, red dot design awards and design fair catalogues, e.g. Milan design fair) has necessarily influenced the results attained. Had other objects produced in the same geographic spaces been considered, different results probably would have been found. An underlying hypothesis for the approach deployed in this chapter is that cultural influences are capital even when not consciously considered by designers, and are hence reflected in the design production itself. A possibly existing design identity and its continuity over time was sought, in order to recognize characteristics and similarities among products. Design production was not only examined within the Lusophone space, but its international acceptance and appreciation was also considered, so that, through this analysis, it would be possible to recognize the character and contours of the design culture in order to give continuity to a tradition of centuries. It then became imperative to perform a new product search to investigate the differentiation against highly visible design traditions, as is the case of Scandinavian, Italian and German designs. Finally, and from the analytical treatment performed to the data collected in the survey mentioned above, two design concepts are presented which combine Portuguese and Lusophone design tradition, respectively, with contemporary materials and technologies. Ultimately, the aim of these concepts was to establish an alliance between the cultural backgrounds of Portuguese design with the numerous technological possibilities that are presented everyday and that enable the achievement of product improvements at various levels. These improvements focus on aspects such as product performance and increasing the quality of human life.

2.2 Product characteristics associated with identity

This section seeks identification of various types of associations that support cultural identity and seeks to illustrate them by adopting a historical perspective. The aim is also to seek answers to the question: over time is there a noticeable continuity and perseverance in the appearance of traces, signs (materials, shapes, graphic markings, colours, and so on) in the production of objects within the Lusophone space, and Portugal? To answer this question a web based search for products with origins in Portugal and Brazil was carried out. As was observed throughout the many examples encountered in our review, over time there is a continuity and perseverance in the appearance of traces, signs (materials, shapes, graphic markings, colours, and other) in the production of objects within the Lusophone space (represented only by Brazil), and Portugal. With regard to the continuity of Portuguese design, analyzing the set of iconic products encountered (regarding the similarities at technical and conceptual levels) the colours that are most used are white, beige, black, green, metallic grey, red, brown and yellow. The materials most used are ceramic, wood, porcelain, cork, metal and leatherette. In what concerns form, the products are characterized by simplicity, rationality, curved shapes, elegance and convening an organic appeal. Portuguese designers innovate especially in incorporating several features to objects, they take care in choosing the most appropriate and up-to-date material, and the products are usually easy to use and provide great comfort, with no graphic markings.

In terms of the continuity of Brazilian design, analyzing the set of iconic products encountered (regarding the similarities between technical and conceptual qualities), suggests that Brazilian designers seem to show a preference for brown, white, black and green colours. Materials-wise, a higher adherence to wood, plastic, leather and metal is visible. Brazilian products are characterized primarily by simplicity, rational and straight lines, wavy and winding forms. These designs are innovative, incorporate functional improvements and demonstrate savings in the materials used in the objects, while designers select the most recent materials and apply high mutability to their projects. The designers of this nationality do not use graphic markings and inferred ease of use of their products varies between easy to medium. In the following section, it is possible to define an identity for Lusophone design, based on the intersection of Brazilian and Portuguese design characteristics.

2.2.1 Portuguese and Brazilian design

Regarding the possibility of a cultural identity of Portuguese design, one can thus conclude that the most common colours are (described by decreasing frequency): white, black, brown, beige, metallic grey, green, red, and cork yellow. The materials preferably used by Portuguese designers in most objects are clay (pottery), wood, cork, porcelain, plastic, metal and leatherette. Regarding the shape of the products, these are characterized by their simplicity, rationality, curved lines, elegance, organic character, softness and in some cases straightness of lines. This design culture stands out for its innovation in the field of materials, and it also reflects concerns about the ease of use, comfort, very often the addition of new materials and products are aesthetically modern. Surveyed objects are mostly devoid of graphic markings, except for the product brand. Finally, all objects are considered to require between easy and medium ability for their use.

Brazil also shows important similarities between its designers' production, in their choice of colours such as brown, white, black and green, this similarity is clear. They use the most common materials including wood, metal, plastic and leather. The sampled products designed in this nation exhibit similarities among each other such as simplicity, straight lines, rationality, and undulating and sinuous lines. Originality and innovation stand out in the evident concern for comfort, functional improvements, material savings, and conscious selection of materials by Brazilian designers and through the mutability given to their products. The objects are mostly devoid of graphic markings and inferred ease of use varies between large and medium, although most of these products were deemed easy to use.

In identifying a possibly existing identity for design among the Portuguese language countries, albeit it was based only in Portugal and Brazil, the following characteristics were identified: colours mostly used are white, brown, black and green; materials are typically wood, plastic and metal. Moreover, the products are characterized mainly by their simplicity, rationality and straight lines. The designers differentiate themselves by speaking of the choice of material, the comfort they bring to the objects, assigning more than one functionality to their products and at the same time incorporating mutability into their designs. The objects created within the Lusophone space are generally easy to use, and are mostly devoid of graphic markings.

2.2.1.1 Sampled Portuguese designs

Besides the 46 product designs showed in this section, an additional set of 26 other products was analyzed in this study, but are not shown due to space and size restrictions (Fig. 4; images shown are in the public domain; for a complete set of references see Simão & Coelho, 2011).

Fig. 4. Images of Portuguese designed products sampled as a basis for analysis.

2.2.1.2 Sampled Brazilian designs

Besides the 32 examples of product design from Brazil shown in this section, an additional set of another 32 products was considered in the analysis presented in this study, but are not shown due to space and size restrictions (Fig. 5; images shown are in the public domain; for a complete set of references see Simão & Coelho, 2011).

Fig. 5. Images of Brazilian designed products sampled as a basis for analysis.

2.2.2 Comparison with Scandinavian, Italian and German design

This section is intended to achieve the objective of identifying the characteristics of Lusophone design identity in the context of other geographically based design traditions, as a form of visible differentiation. Hence, it seeks to identify differences between the products of Lusophone origin and products with a Scandinavian, Italian and German origin.

As shown in this section, there are some differences between Lusophone design and Scandinavian, Italian and German design. This section enables establishing material and use based differences drawn from the four design origins included in the study.

With regard to colour preference very significant differences do not exist, however, Lusophone design resembles Scandinavian design in this respect, differing from Italian and German design by the use of more subtle and neutral colours. The colours that are primarily used by the Italian current tend to be more flashy (Table 1).

Lusophone Space (136)		Scandinavia (23)		Italy (26)		Germany (23)	
White	25%	White	35%	White	36%	Black	36%
Brown	17%	Red	30%	Metallic Grey	32%	Metallic Grey	28%
Black	15%	Black	30%	Yellow	24%	White	28%
Green	7%	Brown	26%	Black	20%	Blue	16%
				Red	20%	Grey	12%
				Blue	16%	Orange	12%
				Pink	12%		
				Brown	12%		
				Green	12%		
				Orange	12%		

Table 1. Colour characteristics prevalent across the sampled products.

In relation to the material differences visible in the material of choice for products, these are shown in Table 2. Portuguese speaking designers have a special preference for wood primarily, followed by plastic, while the materials of preference of Scandinavian, Italian and German designers (metal) are the least utilized by Lusophone designers.

Lusophone Space (136)		Scandinavia (23)		Italy (26)		Germany (23)	
Wood	17%	Metals	48%	Metals	52%	Metals	32%
Plastics	6%	Plastics	30%	Plastics	40%	Wood	24%
Metals	4%	Wood	26%	Wood	24%	Leather	20%
		Fabric	17%			Plastics	16%
		Glass	17%				

Table 2. Materials that are prevalent across the sampled products.

At a formal level, design projects with Lusophone and German origins display a great sobriety instilled in the shape of products, while designs from Italy and Scandinavia display more organic and fun shapes than those from Germany and the Lusophone space (Table 3). Innovation in the Lusophone space is still lagging behind the other design streams examined. Although Lusophone products reflect innovation and originality, they are still short of the originality that grew with these other three schools for decades and contributes to highlighting the timeless tradition of their designs (Table 4). Across the items displayed in Table 5 there is not much difference, since the products of the four nationalities and, or, regions, are usually devoid of graphic markings, using them only to show the product's brand. Products are mostly similar in terms of inferred ease of use (Table 5).

Lusophone S. (136)		Scandinavia (23)		Italy (26)		Germany (23)	
Simplicity	21%	Simplicity	52%	Simplicity	32%	Simplicity	52%
Rationality	14%	Rationality	26%	Round Lines	20%	Minimalism	44%
Straight Lines	5%	Organic Shapes	13%	Fun Shapes	16%		
				Funcionality	12%		

Table 3. Form characteristics that predominate in the products sampled.

Lusophone S. (136)		Scandinavia (23)		Italy (26)		Germany (23)	
Changeable	6%	Adaptable	13%	Innovative Technology	16%	Adaptable	28%
Materials	5%	Eco-Sustainable	13%	Design classics	12%	Innovative Technology	24%
Comfort	5%	Modern	13%	Compact	12%	Comfort	20%
		Multiple functions	13%	Fun Shape	12%	Multiple functions	16%
				Multiple functions	12%	Modular	12%
				Ergonomic	12%		
				Changeable	12%		

Table 4. Characteristics of originality and innovation prevalent across the sampled products.

Lusophone S. (136)		Scandinavia (23)		Italy (26)		Germany (23)	
Graphical Markings							
Devoid	96%	Devoid	100%	Devoid	80%	Devoid	76%
Brand	4%			Brand	20%	Brand	24%
Perceived Ease of Use							
Easy	96%	Easy	100%	Easy	88%	Easy	88%
Average	4%			Average	12%	Average	12%

Table 5. Prevalent characteristics of ease of use and the presence of graphical markings in the products sampled.

The analysis presented in this section suggests that Lusophone design shows some differences when compared to Scandinavian, Italian and German design traditions, particularly in relation to innovation, which is rather less inculcated in Portuguese and Brazilian products. This is deemed to result largely from the tradition and heritage that comes from long ago in these design currents. There are also obvious similarities that unite these four design streams, namely at the form level.

2.2.2.1 Sampled Scandinavian designs

The sample consists of 23 product designs, which are the basis on which the analyses relating to Scandinavian design are made in this study (Fig. 6; images shown are in the public domain; for a complete set of references see Simão & Coelho, 2011).

Fig. 6. Images of Scandinavian designed products sampled as a basis for analysis.

2.2.2.2 Sampled Italian designs

The sample includes 26 products, designed both by Italian and other designers commissioned by Italian companies, for products sold as Italian products (Fig. 7; images shown are in the public domain; for a complete set of references see Simão & Coelho, 2011).

Fig. 7. Images of Italian designed products sampled as a basis for analysis.

2.2.2.3 Sampled German designs

The sample consists of 23 product designs, which are the basis on which the analyses relating to German design are made in this study (Fig. 8; images shown are in the public domain; for a complete set of references see Simão & Coelho, 2011).

Fig. 8. Images of German designed products sampled as a basis for analysis.

2.3 Discussion on the use of product profiles to generate new concepts

This section discusses the design concepts proposed in continuity to the characteristics found, but adopting contemporary or emerging technology and materials. Two concepts were proposed (Figures 2 and 3). One was designed taking into account the characteristics of Portuguese products taken from the analysis done for Portuguese products. The other one reflects the characteristics of Lusophone joint design identity. These concepts seek to provide continuity to the two design cultures focused, through the selection of factors which were set similarly to the existing sampled products. These factors include the colours most frequently used by designers of these nationalities, their chosen materials and the formal characteristics of their products. Innovation was sought in these creative concepts, in order to distinguish these from existing products on the market. The focus of the first concept fell on power savings, i.e. on an economic level, without neglecting the functional level (Figure 2). The innovations inculcated in the second concept concern mainly the formal domain, in an attempt to make the product both functional and versatile, and in such, conferring adaptability to satisfy changing and dynamic user needs (Figure 3).

The results suggest that, in order to continue a tradition of centuries without which the designed products will no longer be accepted within and outside the Lusophone space, these should incorporate colours, materials and forms typical of the Portuguese and Lusophone culture. Colours of choice of Portuguese and Lusophone designers, identified as a result of the analysis undertaken in this study, are white, brown, beige, green, metallic grey, red, cork colour tones, yellow and blue. The materials selected by these designers are usually wood, ceramic, cork, plastic, porcelain, metal, steel, aluminium, and vinyl or leather. At a formal level, the products reflect simplicity, rationality, curved lines, elegance, organic character, smoothness and straightness of lines. Designs should also reflect increasing concerns for sustainability, ecological values and advanced functional, since the products designed by designers of these nationalities are mostly very easy to use and should offer more consistently clean and sustainable solutions to problems faced by the consumer society in the current times.

Significant similarities were found between the design productions sampled in this study. Portuguese design production, as sampled in this study, shows a preference for colours such as white, black, beige, brown and metallic grey. In what concerns materials, the choice falls mainly on ceramics, wood and cork; in terms of shape or form, products are simple, rational and often incorporate curved lines. Although Portuguese product designs show a striking low level of innovation, designed products are deemed easy to use and are mostly devoid of graphical markings. For Brazilian designers, it can be concluded that they prefer colours like brown and white, in terms of materials, their preference falls on wood and in terms of forms, their products are conspicuously simple. Brazilian designers innovate in particular in products that integrate technology and that are comfortable, while Portuguese designers innovate mostly by conceiving products that are very user-friendly. Portuguese designs are mostly devoid of graphical markings. From the joint analysis of the sampled designs pertaining to these two nationalities it can be concluded that Lusophone design gives primacy to colours like white, brown and black; wood is the material of choice and the form of these products is simple and rational. The rate of innovation in Lusophone product design is not high, but designers produce user-friendly products which are devoid of graphical markings.

As a result of the analysis presented, Scandinavian chromatic preferences reflect mainly white, red, black and brown. In respect to the materials' order of preference, it begins with plastic, followed by wood, metal, textiles, glass and, finally, their products are also characterized by simplicity, rationality and the use of organic forms. The innovations incorporated in these are evident at the level of adaptability, sustainability, multi-functionality of products and modern appearance. The objects designed within this culture are devoid of graphic markings and are very easy to use. Italian design uses more often as colours white, metallic grey, yellow, black, red, blue, pink, brown, green and orange. In terms of materials that stand out, there is plastic, metal, with special focus on steel, and wood. At the form level, products are characterized by simplicity, curved shapes, fun shapes and functional form. Their originality can be seen through the adaptation of new technologies to design, which led to the creation of great classics of design, striving to create compact objects. Multi-functionality, ergonomics, and fun are common product attributes. Italian product designs include some graphic markings although most products are devoid of them. Inferred ease of use ranges from medium to easy. Finally, in what concerns the German current, based on the sampled designs covered within this study, often designers opt for black, metallic grey, white, blue, grey and orange colours. In terms of materials there is a preference for wood, leather, steel and plastic. These products' main characteristics are simplicity and minimalism. Their originality lies on adaptability and incorporation of new technology, great comfort, modularity and multiple functionality. The products designed in Germany are mostly devoid of graphical markings and if they do have them, they concern the product brand. These products are deemed mostly easy to use.

In this era of globalization accelerated by technology, although it is not noticeable at first glance, there seem to be apparent marks of national design in the existing design production, even if a conscious effort to create them was absent from the design process. It is a fact that the design originating in different nationalities and cultures is similar in many ways, but the designer, is influenced by culture, societal norms and environmental conditions of the place where he or she grows and matures. Therefore, even if there is no deliberate intent, design will always reflect personal characteristics and the experiences of those who design the products, even if sometimes barely visible.

3. Mediation by product personalities to transfer Portuguese and Lusophone cultural traits to product design

The approach reported in the second part of this chapter seeks to explicitly identify cultural traits, and tentatively embed a selection of these in the design of products, in order to propose a direct method to confer an interpreted cultural identity to products undergoing the process of design. Hence, positive and neutral cultural traits were selected, after identifying the features of the cultural identities focused (study I). Thus, the application of the methodology began with unveiling the Portuguese positive and neutral traits and the commonalities between the positive and neutral identity aspects within the Lusophone cultural identities. Based on these features, another study (study II) was conducted to match these cultural traits with the personality attributes of the product.

After matching the selected cultural traits with the personality dimensions of the product, a further study was conducted (study III), by selecting, as examples, a set of clothes pressing warm irons (4) and a set of coffee machines (8). These were examined with respect to a listing of the 17 personality dimensions, and considering the matching of basic technical characteristics for each product to its position and placement personality-wise.

Another study (study IV), taking into account the previous match, was carried out establishing the relationship between personality attributes and technical characteristics of the products tested in the previous study (study III). From this process, two product profiles were obtained as a result, which were then implemented in two product lines, a Portuguese and a Lusophone one. These product line results were chosen from a broad base of concepts generated, considering objective criteria. After the generation of concepts for the two product lines, there was an empirical validation by sampling of industrial design students (study V) to confirm whether the proposals developed did turn out to reflect Portuguese cultural identity and Lusophone cultural identity, respectively.

3.1 Method deployed to transfer cultural traits to product requirements

The development of the second project reported in this chapter was structured by a methodology that sought to satisfy an overarching aim and specific goals and provide answers to their inherent research questions. The overarching aim was defined as identifying the aspects that define Portuguese and Lusophone cultural identities, adopting both an internal and an external perspective, and seeking to extrapolate these cultural identity traits, in order to contribute to develop a Portuguese design identity (for Lusophone consumption) and a Lusophone design identity (for global consumption).

One specific goal was set as 'performing a survey of Portuguese and Lusophone identity traits, adopting a cultural perspective'. Another one was defined as 'translating the cultural traits identified, in a positivist approach, to a product line with Portuguese character and to a product line with Lusophone character'. The research questions that guided the development of the project were:

- What are the collective cultural identity traits of the Portuguese and Lusophone cultures (seen from the inside and from the outside)? (study I)
- From the set of identified cultural traits, which of these may be regarded as positive and neutral in order to be inculcated in the design production? (study I)
- Is the assignment of product personalities a suitable means of transferring cultural traits into product qualities? (studies II, III, IV and V).

3.2 Study I – Portuguese and Lusophone cultural traits

The study reported in this section, concerning cultural inquiry, was based on literature review to unveil a set of opinions from respected scholars within the humanities disciplines (sociology, anthropology, philosophy) and the relational study of some areas of arts and fine arts. Rather than an exhaustive listing of the whole set of cultural traits surveyed, a subset of results is presented. Partial results obtained for study I are shown in Tables 6 and 7, for

Adventurer (history and humanities) Audacious (literature) Autognose (history and humanities, literature) Autonomous (history and humanities, literature) Aware (history and humanities) Bold (history and humanities, literature) Bucolic (history and humanities) Concrete (cinema, history and humanities) Confident (history and humanities) Contrast (painting) Eclecticism (painting) Colourful (song) Cosmopolitan (history and humanities, painting) Dichotomy aesthetics (paint) Different (history and humanities) Disseminator (cinema) Diverse (cinema, history and humanities, painting) Dreamer (history and humanities) Ductile (history and humanities)	Dynamic (history and humanities, painting) Empirical (history and humanities) Enthusiast (cinema, history and humanities) Epic (history and humanities, literature, music) Ethical (history and humanities) Experimental (cinema) Overview autotelic (cinema) Experimental aestheticism (cinema) Expressive (painting) Flash (history and humanities) Golden (history and humanities) Gracious (history and humanities) Harmony (history and humanities, music) Hetero-textual (literature) Hope (history and humanities, music) Ideological (cinema, history and humanities, painting) Imaginative (history and humanities, painting)	Independent (literature) Industrious (history and humanities) Intellectual Property (history and humanities) Liberal (history and humanities, painting) Likely (history and humanities, literature) Lucid (history and humanities) Modest (history and humanities) Multi-mode (history and humanities) Naturalistic (literature, painting) Noble (history and humanities) Organic (history and humanities) Organized (history and humanities) Original (cinema, history and humanities, literature) Paradigmatic (cinema, history and humanities) Picturesque (history and humanities, painting) Prodigious (history and humanities) Rationalist (history and humanities)	Realistic (cinema, history and humanities, literature) Respectful (history and humanities) Search (literature) Self-consciousness (history and humanities) Self-critical (cinema) Self-reflection (cinema) Sensible (history and humanities) Solidarity (history and humanities) Spontaneous (history and humanities) Strategic Intelligence (history and humanities) Suave (history and humanities) Sublimation (history and humanities) Subtle (history and humanities) Tolerant (history and humanities) Universal (history and humanities) Unmistakable (history and humanities) Virtue (painting) Vital (painting, music)

Sources: Almeida (1995), Bello (2009), Cabral (2003), Castagna (2005), Costa (1998), Lemière (2006), Lourenço (1994, 2001), Moreira (2005), Neto (2005), Quadros (1999), Rodrigues & Devezas (2009).

Table 6. Cultural aspects with a positive nature concerning Portugal (in parentheses the thematic track of the literature review from which the cultural trait was retrieved is indicated).

Abstract (painting)	Elusive (history and humanities)	Naive (history and humanities)
Acumen (literature)	Emblematic (literature, painting),	Needy (history and humanities)
Adaptive (history and humanities)	Feeling depth (history and humanities)	Nostalgia (history and humanities, literature, music)
Allegory moralizing (painting)	Fey (history and humanities)	Ornamental (painting)
Ambiguous (history and humanities)	Fini-secular (literature)	Pantheistic (history and humanities)
Antagonist (history and humanities, literature)	Folklore (history and humanities)	Parental (history and humanities)
Aseptic (history and humanities)	Heroic (history and humanities, literature)	Patriotic (cinema, history and humanities)
Belief in miracles (history and humanities)	Hidden (history and humanities)	People (cinema, history and humanities)
Buck (history and humanities)	Honour (literature)	Proud (literature)
Candor (history and humanities)	Humble (history and humanities)	Radical (history and humanities)
Centred (humanistic-historical)	Hybrid (literature)	Romantic (history and humanities, literature)
Christian (history and humanities, literature)	Hyper-identity (history and humanities)	Allogeneic (history and humanities)
Collective (history and humanities)	Idyllic (history and humanities)	Sacred (literature)
Complex (history and humanities)	Imperial (history and humanities)	Sacrificed (history and humanities)
Concentrate (history and humanities)	Improvisation (history and humanities)	Sadness (history and humanities, music)
Constant (history and humanities)	Incremental (historical and humanistic)	Sensitive (history and humanities)
Controllable (history and humanities)	Intense religiosity (cinema, painting)	Sentimental (history and humanities, literature, music)
Creator (history and humanities)	Interstitial (literature)	Single (history and humanities, painting)
Critical (history and humanities)	lyric (literature)	Singular (cinema, history and humanities)
Cultism (literature) Catholic (painting)	Metamorphic (history and humanities)	Spiritual (history and humanities)
Cultural assimilation (history and humanities)	Militant (history and humanities)	Stubborn (history and humanities)
Curvilinear reasoning (history and humanities)	Mimetic (history and humanities)	Subjective (literature)
Diachronic (cinema)	Miscegenation (literature)	Subversive (cinema)
Diaspora (history and humanities)	Moral (painting) Ancient (story)	Sync (cinema) Theology (history and humanities, literature)
Dogmatic (history painting)	Morphological (history and humanities)	Utopian (history and humanities)
dream Themes (painting)	Movement (history and humanities)	Water (history and humanities)
Ecumenical (history and humanities)	Mystery (history and humanities, music)	

Sources: Almeida (1995), Baguet (1999), Bello (2009), Borja (2005), Cabral (2003), Cademartori (2006), Cardoso & Catelli (2009), Castagna (2005), Costa (1998), Domingues (2000), Grosso (1999), Lemière (2006), Lourenço (1994, 2001), Martins, Sousa & Cabecinhas (2006), Matos-Cruz (1999), Moreira (2005), Nascimento (2009), Neto (2005), Netto, Dias & Mello (2003), Ngai (1999), Ono (2004), Pereira (1999), Quadros (1999), Rago (2006), Ribeiro 82004), Rodrigues & Devezas (2009), Rossini (2005), Salvo (2006), Silva (1999).

Table 7. Cultural aspects with a neutral nature concerning the Lusophone space (in parentheses the thematic track of the literature review from which the cultural trait was retrieved is indicated).

Portuguese positive aspects and Lusophone neutral aspects (the distinction between positive, neutral and negative aspects was done by the authors).

3.3 Study II - Matching selected cultural traits with product personality dimensions

The cultural traits obtained from study I were corresponded by the authors to Jordan's (2000) product personality attributes. Each cultural trait was assigned to one or more of the product personality dimensions (Table 8) and a matrix was prepared that translated the cultural traits into personality dimensions. The personality dimensions that resulted are presented in Tables 9 (results of subjective transfer of the Portuguese cultural traits identified in study I) and 10 (results of subjective transfer of the Lusophone cultural traits identified in study I).

kind - somewhat kind - neither kind or unkind - somewhat unkind - unkind
honest - somewhat honest - neither honest or dishonest - somewhat dishonest - dishonest
serious minded - somewhat serious minded - neither serious minded or light hearted - somewhat light hearted - light hearted
bright - somewhat bright - neither bright or dim - somewhat dim - dim
stable - somewhat stable - neither stable or unstable - somewhat unstable - unstable
narcissist - somewhat narcissist - neither narcissist or humble - somewhat humble - humble
flexible - somewhat flexible - neither flexible or inflexible- somewhat inflexible - inflexible
authoritarian - somewhat authoritarian - neither authoritarian or liberal - somewhat liberal - liberal
driven by values - somewhat driven by values - neutral - somewhat not driven by values - not driven by values
extrovert - somewhat extrovert - neither extrovert or introvert - somewhat introvert - introvert
naïve - somewhat naïve - neither naïve or cynical - somewhat cynical - cynical
excessive - somewhat excessive - neither excessive or moderate - somewhat moderate - moderate
conforming - somewhat conforming - neither conforming or rebellious - somewhat rebellious - rebellious
energetic - somewhat energetic - neither energetic or non energetic - somewhat non energetic - non energetic
violent - somewhat violent - neither violent or gentle - somewhat gentle - gentle
complex - somewhat complex - neither complex or simple - somewhat simple - simple
optimist - somewhat optimist - somewhat pessimist - pessimist

Table 8. Product personality dimensions (Jordan 2000).

Upper product personality attribute	Lower product personality attribute
Kind	Neither kind or unkind
Honest	Somewhat dishonest
Somewhat fun	Somewhat serious
Bright	Somewhat dim
Stable	Somewhat unstable
Humble	Neither humble or narcissistic
Flexible	Inflexible
Liberal	Authoritarian
Driven by values	Somewhat driven by values
Somewhat extroverted	Somewhat extroverted
Naïve	Somewhat cynical
Moderate	Excessive
Somewhat conforming	Somewhat rebellious
Energetic	Somewhat energetic
Gentle	Somewhat violent
Simple	Complex
Optimistic	Somewhat pessimistic

Table 9. Product personality attribute ranges resulting from translating the Portuguese cultural traits identified in study I (transfer performed by the authors).

Upper product personality attribute	Lower product personality attribute
Kind	Neither kind or unkind
Honest	Somewhat dishonest
Somewhat fun	Somewhat serious
Somewhat bright	Somewhat dim
Stable	Somewhat unstable
Humble	Neither humble or narcissistic
Flexible	Inflexible
Liberal	Somewhat liberal
Driven by values	Somewhat not driven by values
Somewhat extroverted	Somewhat introverted
Naïf	Somewhat cynical
Moderate	Excessive
Neither conforming or rebellious	Somewhat rebellious
Energetic	Somewhat energetic
Gentle	Somewhat gentle
Simple	Complex
Optimistic	Neither optimistic or pessimistic

Table 10. Product personality attribute ranges resulting from translating the Lusophone cultural traits identified in study I (transfer performed by the authors).

3.4 Study III – Correspondence of product personality dimensions to product attributes

Some examples of objects comprised of four clothes pressing irons and eight coffee machines were chosen (Fig. 9), in order to make an analysis of these objects with regard to the Product Personality Assignment technique by Patrick W. Jordan (2000). The assignment of personality attributes was carried out by a panel of eight third year undergraduate industrial design students (aged from 20 to 23 years old) that rated each object in terms of the personality dimensions in a 5 point Lickert scale ranging from the personality attribute to its opposite (e.g. kind - unkind) and three intermediate ratings (e.g. somewhat kind, neither kind or unkind, somewhat unkind), according to Table 3. The eight raters analysed the objects grouped in three sets, one of clothes pressing irons and two of coffee machines. The Kendall coefficient of concordance was used to assess the consistency of ratings among the panel.

Fig. 9. Products that were used as a basis for the product personality assignment survey performed as part of study III.

The ranking attained by combining the judgement of the eight raters within the personality pairs of each set of four products is shown in Tables 11 to 13, accompanied by the result of the Kendall coefficient of concordance for each dimension and set rated.

Personality Attribute	Ranking	Personality Attribute	Significance
Kind	B – C – A– D	Unkind	Significant at 95% c.i.*
Honest	C – B – A – D	Dishonest	Not significant
Serious	C – B – D – A	Light-hearted	Significant at 99% c.i.*
Bright	B – C – A – D	Dim	Significant at 95% c.i.*
Stable	C – B – A – D	Unstable	not significant
Narcissistic	C – D – A – B	Humble	Significant at 99% c.i.*
Flexible	B – A – D – C	Inflexible	not significant
Authoritarian	C – D – A – B	Liberal	Significant at 99% c.i.*
Driven by values	C – B – A – D	Not driven by values	Not significant
Extrovert	A – D – B – C	Introvert	Significant at 95% c.i.*
Naïve	B – A – C – D	Cynical	Significant at 95% c.i.*
Excessive	D – C – A – B	Moderate	Significant at 95% c.i.*
Conforming	C – B – D – A	Rebellious	Significant at 95% c.i.*
Energetic	A – B – D – C	Non energetic	Significant at 99% c.i.*
Violent	D – C – A – B	Gentle	Significant at 99% c.i.*
Complex	C – D – B – A	Simple	Not significant
Pessimistic	C – D – B – A	Optimistic	Significant at 99% c.i.*

* - c.i. – confidence interval

Table 11. Aggregate ranking of the four clothes pressing irons depicted in Fig. 9 for each of the 17 personality dimension pairs and calculation of significance of agreement (based on Kendall correlation coefficient, Siegel & Castellan 1988).

Personality Attribute	Ranking	Personality Attribute	Significance
Kind	B – A – C – D	Unkind	Significant at 99% c.i.*
Honest	B – C – A – D	Dishonest	Significant at 99% c.i.*
Serious	B – D – C – A	Light-hearted	Significant at 99% c.i.*
Bright	B – A – C – D	Dim	Not significant
Stable	B – C – A – D	Unstable	Significant at 99% c.i.*
Narcissistic	D – C – A – B	Humble	Significant at 99% c.i.*
Flexible	A – B – C – D	Inflexible	Significant at 95% c.i.*
Authoritarian	D – C – A – B	Liberal	Significant at 95% c.i.*
Driven by values	B – C – D – A	Not driven by values	Significant at 95% c.i.*
Extrovert	A – C – D – B	Introvert	Significant at 99% c.i.*
Naïve	B – A – C – D	Cynical	Significant at 99% c.i.*
Excessive	D – C – A – B	Moderate	Significant at 99% c.i.*
Conforming	B – C – D – A	Rebellious	Significant at 99% c.i.*
Energetic	A – C – D – B	Non energetic	Significant at 95% c.i.*
Violent	D – A – C – B	Gentle	Significant at 95% c.i.*
Complex	D – C – A – B	Simple	Significant at 99% c.i.*
Pessimistic	D – B & C – A	Optimistic	Not significant

* - c.i. – confidence interval

Table 12. Aggregate ranking of the first set of four coffee machines depicted in Fig. 9 for each of the 17 personality dimension pairs and calculation of significance of agreement (based on Kendall correlation coefficient, Siegel & Castellan 1988).

Personality Attribute	Ranking	Personality Attribute	Significance
Kind	D - C - B - A	Unkind	Significant at 99% c.i.*
Honest	B - C - D - A	Dishonest	Significant at 95% c.i.*
Serious	B - A - C - D	Light-hearted	Significant at 99% c.i.*
Bright	B - C - D - A	Dim	Not significant
Stable	B - C - D - A	Unstable	Significant at 99% c.i.*
Narcissistic	A - D - B - C	Humble	Significant at 95% c.i.*
Flexible	D - C - B - A	Inflexible	Significant at 99% c.i.*
Authoritarian	A - B - C - D	Liberal	Significant at 99% c.i.*
Driven by values	B - A - C - D	Not driven by values	Significant at 95% c.i.*
Extrovert	D - C - A - B	Introvert	Significant at 99% c.i.*
Naïve	C - B - D - A	Cynical	Significant at 99% c.i.*
Excessive	A - D - C - B	Moderate	Not significant
Conforming	B - C - A - D	Rebellious	Significant at 99% c.i.*
Energetic	A - D - C - B	Non energetic	Not significant
Violent	A - B - D - C	Gentle	Significant at 99% c.i.*
Complex	A - B - C - D	Simple	Significant at 99% c.i.*
Pessimistic	B - A - C - D	Optimistic	Not significant

* - c.i. – confidence interval

Table 13. Aggregate ranking of the second set of four coffee machines depicted in Fig. 9 for each of the 17 personality dimension pairs and calculation of significance (based on Kendall correlation coefficient, Siegel & Castellan 1988).

Materials	metals	wood	ceramics	plastic	ecological
Colour	primary	pastel	metallic	warm	cold
Shape	straight	organic	coherent	contrasting	functional
Graphic markings	geographical	decorative	instructions	patterns	reliefs
Archetype	conventional / traditional	minimalist	luxury	utilitarian	adaptable
Morphology	single part	few parts	dependency between parts	modularity	many interconnected systems
Ease of use	simple and intuitive	complex, yet intuitive	neither easy or difficult to use	not very complex, but difficult to use	very complex and difficult to use
Production Process	handicraft	rudimentary industrial	contemporary industrial	manufacturing by high technology	user fabrication
Technological Sophistication	moving parts (manual)	electrical technology	electronic technology	information technology	nanotechnology and biotechnology
Multiple functionality	single function	few functions	some functions	various functions	many functions
Size	very small	small	medium	great	very large

Table 14. Product technical dimensions each broken down into five categories, that were considered in study III.

3.5 Study IV – Establishing the link between product personalities and product characteristics

The 12 objects depicted in Fig. 9 were further characterized, by the authors, in terms of their product attributes according to a series of dimensions. These included materials, colour, shape, graphic markings, archetype, morphology, inferred ease of use, manufacturing process, technological sophistication, multiple functionality and size. The dimensions that were used to characterize the 12 objects involved in study III are shown in Table 14.

As a result of study IV, two product attribute lists were attained, one concerning the transference of Portuguese cultural traits to product properties and the other one concerning the transfer of Lusophone cultural traits (Table 15).

Product technical dimension	Culturally induced Portuguese product profile	Culturally induced Lusophone product profile
Colour	Cold	Cold
Shape	Straight, coherent, contrasting	Straight, coherent, contrasting
Graphical markings	Decorative, instructions	Decorative, instructions
Archetype	Minimalist	Minimalist
Multiple functionality	Single function	-
Size	Small	Small
Ease of use	-	Complex, yet intuitive

Table 15. Product attributes attained as a result of study III.

3.6 Study V – Generation of product concepts and their validation

Various living room furniture concepts were generated based on two product specifications that took as starting points the results presented in Table 15 and that were enlarged considering anthropometric (Panero & Zelnik, 2002) and other requirements. These initial concept sketches were evaluated by the authors, with respect to criteria derived from the specification and were also subjected to the scrutiny of 21 second year undergraduate industrial design students (aged from 19 to 22 years old). These did not however show significant agreement in terms of their preference among the concepts generated. The authors' evaluation matrix (based on an expanded requirements list developed within the design process) led to the detailed development of the concepts depicted in Figures 10 and 11, respectively, a product line based on the Portuguese cultural traits, named "Vale", and one based on the Lusophone ones, named "Império".

In order to test the effectiveness of the approach reported in this chapter, the respondents in this study were asked to identify, from the concepts generated, which of those had either Portuguese traits, Lusophone traits or none. These results are shown in Table 16.

Furniture concept	Portuguese traits	Lusophone traits	No Portuguese or Lusophone traits
"Império" - Lusophone A	8	8	5
"Bloco" - Lusophone B	3	9	9
"Flex 2" - Lusophone C	2	4	15
"Vale" - Portuguese A	5	6	10
"Leveza" - Portuguese B	7	7	7
"Flex" - Portuguese C	2	8	11

Table 16. Survey seeking the validation of the results of the studies reported in this chapter (21 respondents - second year undergraduate industrial design students).

The results of the survey do not lead to consider that the results convey clearly identifiable cultural traits associated with the Portuguese and Lusophone cultures, although these vary from product concept to product concept.

Fig. 10. Renders of "Vale" living room furniture line based on the Portuguese cultural traits and their corresponding product technical attributes (designed by the second author).

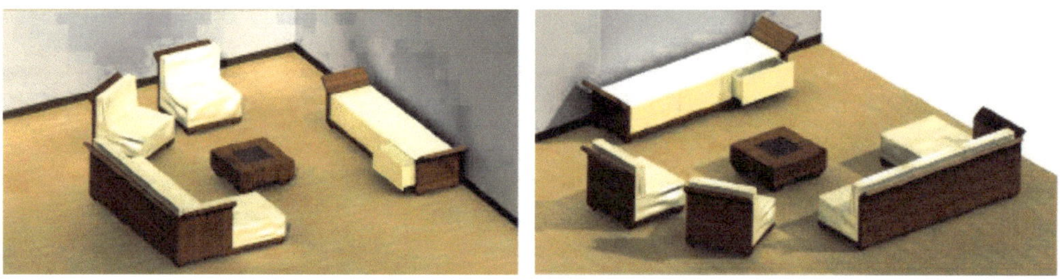

Fig. 11. Renders of "Império" living room furniture line based on the Lusophone cultural traits and their corresponding product technical attributes (designed by the second author).

3.7 Discussion of the results of the five studies presented

In the second part of this chapter, cultural traits were the starting point to reach at the product profiles that were used as the basis for the design of two furniture lines. The scope of the work reported is not limited to furniture and is deemed applicable in a wider scope, considering its genesis and methodology, based on a literature review of cultural traits, taking into account the personalities of consumer products and consulting industrial design students.

Advancing the knowledge on the transfer of cultural traits to product design features may require further inquiry, since the projects reported in this chapter did not lead to conclusive results towards accepting or rejecting the hypothesis which was presented in the methods section of this chapter. The adequateness of the use of the product personality assignment technique in supporting this transfer could not be determined conclusively, as the results of the panel convened to assess the cultural identity of the product concepts produced was not conclusive, lacking agreement among the group.

4. Conclusion

One of the underlying assumptions for this chapter was that there is a space yet to be filled by a culturally inspired design current to take the space of design production with a Lusophone flavour, for international dissemination. Another underlying assumption is that design may be inspired by culture, and a direct transfer of cultural traits to product attributes may be pursued, with the support of product personality attributes, as a mediator. The results of both streams of analysis (Fig. 1) presented in this chapter were found not to match entirely. The product feature profiles encountered in the sampled Lusophone design production (first part of the chapter) do not match entirely the product feature profiles obtained from transferring selected cultural traits (positive and neutral) to technical features. This suggests that there is a space yet to be filled by a culturally inspired design current to take the space of design production with a Lusophone flavour, for international dissemination. This mismatch also suggests that culturally inspired design may yield novel design concepts and trigger an array of new proposals that cater to varying tastes and sensibilities. This may contribute to more visibility of designs from particular regions of the globe, opposing the paradigm of mainstream design production, according to the traditional and well established design schools and currents. In a globalized world, culturally inspired design carries the promise to bring more cultural inclusion into the design discourse and to promote added differentiation of consumer choice of design objects.

5. Acknowledgment

The research projects presented in this chapter were developed as Master of Science thesis in industrial design engineering by the second and third authors, supervised by the first author. A selection of results from these projects have previously appeared in Simão and Coelho (2011) and Silva and Coelho (2011), published by Common Ground.

6. References

Almeida, Onésimo Teotónio. (1995). Em busca da clarificação do conceito de identidade cultural. O caso açoriano como cobaia. in Separata de *A Autonomia no Plano Sócio-Cultural.* Congresso do I Centenário da Autonomia dos Açores. Ponta Delgada, Jornal de Cultura, Vol. II.

Baguet Jr., Gabriel. (1999). Percursos e trajectórias de uma historia: A música em Macau na transição de poderes. *Revista Camões* nº07, Macau, p. 84. (URL: http://cvc.instituto-camoes.pt/conhecer/biblioteca-digital-camoes/cat_view/62-revistas-e-periodicos/69-revista-camoes/907-revista-no07-macau.html?start=10

Bello, Maria do Rosário Lupi. (2009). *Implosão do cinema português: duas faces de uma mesma moeda.* Lisboa: Universidade Aberta, 24 p. (URL: http://repositorioaberto.univ-ab.pt/handle/10400.2/1310).

Borja, Juliana. (2005). *A formação da cultura nacional e seus impactos na cultura das organizações do Brasil.* Faculdade Ruy Barbosa, programa de Iniciação Científica, 8 p. (URL: http://www.frb.br/ciente/2006.1/ADM/ADM.BORJA.pdf).

Cabral, Manuel Villaverde. (2003). *A identidade nacional portuguesa: conteúdo e relevância.* Rio de Janeiro, Vol.46, No.3. (URL: http://www.scielo.br/scielo.php?pid=S0011-52582003000300004&script=sci_arttext&tlng=pt).

Cademartori, Nill. (2006). *A cultura como um fator determinante no modo de administrar e de fazer negócios.* 14 p. (URL: http://www.administradores.com.br/informe-se/producao-academica/a-cultura-como-um-fator-determinante-no-modo-de-administrar-e-de-fazer-negocios/129/

Cardoso, Shirley Pereira & Catelli, Rosana Elisa (2009). *O cinema brasileiro dos anos 90 e 2000: Alguns apontamentos presentes na bibliografia contemporânea.* Universidade Estadual de Santa Cruz (UESC), 15 p.

Castagna, Paulo. (1995). Musicologia portuguesa e brasileira: a inevitável integração. *Revista da Sociedade Brasileira de Musicologia,* São Paulo, No.1, pp.64-79. (URL: http://people.ufpr.br/~budasz/CD07.pdf).

Costa, Daciano da. (1998). Design e Mal-Estar. *Colecção "Design, Tecnologia e Gestão".* Porto: Centro Português de Design. ISBN 972-9445-07-9.

Domingues, Maria Manuela A. B. (2000). *Estratégias femininas entre as bideiras de Bissau.* Universidade Nova de Lisboa, F.C.S.H., 529 p. (URL: http://purl.pt/5467/1/sc-91515-v_PDF/sc-91515-v_PDF_X-C/sc-91515-v_0000_1_tX-C.pdf

Grosso, Maria José Reis. (1999). Macau, identidade multilingue. *Revista Camões* nº07, Macau, pp. 96-101. (URL: http://cvc.instituto-camoes.pt/conhecer/biblioteca-digital-camoes/cat_view/62-revistas-e-periodicos/69-revista-camoes/907-revista-no07-macau.html?start=10)

Hofstede, Geert (1980). *Culture's Consequences: International Differences inWork-related Values.* Sage, London.

Jordan, Patrick W. (2000). *Designing Pleasurable Products: an introduction to the New Human Factors.* London: Taylor & Francis, 216 p.

Julier, Guy (2007). *The culture of Design.* Sage publications, 2nd ed., 256 pp.

Lemière, Jacques. (2006). «Um centro na margem»: o caso do cinema português. *Análise Social,* Vol. XLI (180), pp.731-765. (URL: http://www.scielo.oces.mctes.pt/pdf/aso/n180/n180a03.pdf).

Lourenço, Eduardo. (1994). *Nós e a Europa ou as Duas Razões.* Lisboa: Imprensa Nacional - Casa da Moeda.

Lourenço, Eduardo. (2001). *Labirinto da Saudade: Psicanálise Mítica do Destino Português.* Lisboa: Gradiva.

Martins, Moisés de Lemos, Sousa, Helena & Cabecinhas, Rosa. (2006). Comunicação e lusofonia: para uma abordagem crítica da cultura e dos média. Colecção: *Comunicação e sociedade;* 6, Porto: Campo das Letras.

Matos-Cruz, José de. (1999). Macau e o cinema. *Revista Camões* nº07, Macau, p. 198. (URL: http://cvc.instituto-camoes.pt/conhecer/biblioteca-digital-camoes/cat_view/62-revistas-e-periodicos/69-revista-camoes/907-revista-no07-macau.html?start=10)

Moreira, Fernando Alberto Torres (2005) *Identidade Cultural Portuguesa: espaço de autonomia e diversidade.* (URL: http://alfarrabio.di.uminho.pt/vercial/zips/moreira2.rtf)

Nascimento, Augusto. (2009). Lusofonia, que Perspectivas Culturais? *Ciclo de Conferências Encontros com a História.* Maputo. Maio de 2009, 20 p. (URL: http://cvc.instituto-camoes.pt/component/docman/doc_details.html?aut=2094)

Neto, Fernando Ferreira da Cunha. (2005). *Recriações de traços identitários da cultura portuguesa nas obras de Eça de Queirós e Fernando Pessoa – A ilustre casa de Ramires e Mensagem.* Universidade Federal de Minas Gerais, 300 p. (URL: http://dspace.lcc.ufmg.br/dspace/handle/1843/ECAP-7QHHJR).

Netto, Alberto Barella, Dias, Emerson de Paulo & Mello, Paulo César R. de. (2003). Análise da obra cultura organizacional e cultura brasileira organizada por Fernando C. Prestes Motta e Miguel P. Caldas. *Revista Electrónica de Administração,* Facef, Vol.02, Edição 02, Janeiro - Junho 2003. (URL: http://www.facef.br/rea/edicao02/ed02_art04.pdf).

Ngai, Gary. (1999). A Questão da Identidade Cultural de Macau. *Revista Camões* nº07, Macau, pp. 46-56. (URL: http://cvc.instituto-camoes.pt/conhecer/biblioteca-digital-camoes/cat_view/62-revistas-e-periodicos/69-revista-camoes/907-revista-no07-macau.html)

Ono, Maristela Misuko. (2004). *Design, cultura e identidade, no contexto da globalização.* Universidade do Estado da Bahia. Salvador, Brasil. Revista Design em Foco, vol.I, no.1. Julho-Dezembro 2004, p. 54. (URL: http://redalyc.uaemex.mx/redalyc/pdf/661/66110107.pdf)

Panero, Julius & Zelnik, Martin. (2002). *Dimensionamento humano para espaços interiores.* Editorial Gustavo Gili, SA, Barcelona, 2002. ISBN: 84-252-1835-7. pp.134-137.

Pereira, Fernando António Baptista. (1999). Uma identidade plural: acerca das artes em Macau. *Revista Camões* nº07, Macau, pp. 58-68. (URL: http://cvc.instituto-camoes.pt/conhecer/biblioteca-digital-camoes/cat_view/62-revistas-e-periodicos/69-revista-camoes/907-revista-no07-macau.html?start=20)

Quadros, António. (1999). *Portugal razão e mistério-livro I: Uma arqueologia da tradição portuguesa,* Lisboa: Guimarães Editores.

Rago, Margareth. (2006). Sexualidade e identidade na historiografia brasileira. Revista *Aulas,* Dossier Identidades Nacionais, No.2, Outubro/Novembro 2006, pp.11-27 (URL: http://www.unicamp.br/~aulas/volume02/pdfs/sexualidade_2.pdf)

Ribeiro, Margarida Calafate. (2004). *Uma História de Regressos – Império, Guerra Colonial e Pós-colonialismo.* Porto: Editorial Afrontamento.

Rodrigues, Jorge Nascimento & Devezas, Tessaleno. (2009). *Portugal – O pioneiro da globalização: A herança das descobertas.* V. N. Famalicão: Centro Atlântico. Julho, 2009, p.502.

Rossini, Miriam de Souza. (2005) O cinema da busca: discursos sobre identidades culturais no cinema brasileiro dos anos 90. Porto Alegre, Revista *FAMECOS,* quadrimestral, No.27, Agosto 2005, pp. 96-104. (URL: http://www.revistas.univerciencia.org/index.php/famecos/article/viewFile/441/368).

Salvo, Fernanda. (2006). Cinema brasileiro da retomada: da pobreza à violência na tela. Revista *Espcom,* Ano 1, No.1, Maio 2006. (URL: http://www.fafich.ufmg.br/~espcom/revista/numerol/ArtigoFernandaSalvo.html).

Siegel, Sidney & Castellan, N. John (1988). *Nonparametric Statistics for the Behavioural Sciences,* New York: McGraw-Hill.

Silva, Ana Sofia da Cunha & Coelho, Denis A. (2011). Transfering Portuguese and Lusophone Cultural Traits to Product Design. *Design Principles and Practices: An International Journal*, volume 5, issue 1, pp. 145-164.

Silva, Angela Maria Bissoli da. (1999). *Influência da Cultura Brasileira nas Organizações: Raízes e Traços Predominantes*. Faculdade Capixaba de Nova Venécia, 12 p. (URL: http://www.univen.edu.br/revista/n010/INFLUÊNCIA%20DA%20CULTURA%20BRASILEIRA%20NAS%20ORGANIZAÇÕES%20-%20RAIZES%20E%20TRAÇOS%20PREDOMINANTES.pdf).

Simão, Carla Sofia de Matos & Coelho, Denis A. (2011). A Search for the Portuguese Cultural Identity Reflected in the Design of Products. *Design Principles and Practices: An International Journal*, volume 5, issue 3, pp. 171-194.

5

Culturally Inspired Design Education: A Nigerian Case Study

Isah B. Kashim, Sunday R. Ogunduyile and Oluwafemi S. Adelabu
Department of Industrial Design, Federal University of Technology, Akure Nigeria

1. Introduction

Among the countries in Africa, Nigeria took a prominent position when it comes to cultural heritage and creative arts which are manifested in her diverse crafts. The products from these indigenous crafts have sufficiently served the economic needs of the people in the local communities. Since the pre-colonial era, various members of distinct ethnic groups engage in a lot of traditional craft practices in addition to the agrarian occupational engagements. The diversities in the creation of these arts and crafts were used to create strong force that drives the socio-cultural life and economy of the people. These unique artistic traditions thrived within families and guilds of skilled craftsmen in various communities as practiced in the areas of wood carving at Awka, Nupe, Benin; Blacksmithing at Biron, Akwa; Pottery at Dikwa, Abuja, Ilorin, Ipetumodu, Afikpo, Isan-Ekiti, Erusu Akoko and Zaria; Brass smithing and Beadworks at Bida; Bronze casting and sculpture at Ife and Benin; Leather works at Oyo and Kano; Cloth weaving at Ilorin, Iseyin, Okene, Ibadan, Ondo; Cloth dyeing at Oshogbo, Abeokuta, Oyo, Ibadan, Ede, Sokoto, Zaria, Bida and Kano; Mat weaving at Ogotun-Ekiti, Ipetu-Ijesa, Ipoti-Ekiti, Ikeji-Ile, Erin-Ijesa, Efon-Alaye, Ikorodu and Aramoko-Ekiti to mention but few.
Retrospectively, the Yoruba began creating magnificent sculptures in terra cotta between 12th and 14th century. Bronze figures were also made during the 14th and 15th century. The scientific mode of creation started with reproductions in the *cire perdue* loss wax method of bronze casting. The pottery apart from serving as object of storage and cooking was also made in honour of Yoruba deity. Leather works and bead making in Yoruba land are used to decorate crowns won by kings, and other articles such as hats and bags while leathers are pierced together to form designs such as royal leather cushions. The blacksmiths are responsible for the fabrication of tools like hoes, axes, knives, chains and hammers for professional and domestic usage. Calabash are carved and used for storage of foods, drinks and musical rattles. Textile weaving and dyeing with embellishment of colourful patterns and motifs satisfied the local clothing needs. These art works across cultures inculcated a lot of iconographic and mythological delineations that expressed their cultural identity, social values, history and beliefs. The origin of arts and craft is a story within the traditional society and was in response to social change which draws its strength from artists' thoughts, inspirations, speculations, observations, experiences, visions, dreams, culture, environment, myths, fantasies, imaginations and nature (Ahuwan, 1993; Sheba, 1993). The exemplary

creative qualities and skills of the craftsmen were put together effectively in the production of highly functional objects that exhibit exceptional aesthetic aura which satisfied the daily needs of the society.

Culture has been noted to be the totality of all the knowledge and values shared by a society. Hence, this paper fully acknowledged the significance of a Culturally Inspired Design Education as the world transforms from a monolithic culture to one with diversity. This paper essentially captures the evolutionary trends and contemporary issues related to Industrial Design education and practices in Nigeria in relation to its current cultural content. It therefore attempts to examine the effects of home breed Industrial Design education as its affects current graduate performances and professional practices in Nigeria.

2. Industrial design: Contextual definitions

Design is a common term used in many endeavors, such as textile design, graphic design, engineering design, architectural design and all processes of purposeful visual creations which are coordinated together to make a meaningful whole (Ogunduyile, 2007). According to the International Council of Society of Industrial Design (ICSID)

"Design is a creative activity which aim is to establish the multi-faceted quality of objects, processes, services and their systems in whole lifecycles. Therefore, design is the central factor of innovative humanization of technology and the crucial factors of cultural and economic exchange ".

Industrial design historically to date has to do with product design for mass production- an art of imaginative development of manufactured products and product system, which satisfy physical needs. It is the art and science concerned with the conception and creation of machine made products and materials. It is also a creative activity of man which has to do with overall quality and usefulness of a product rather than improving its appearance alone (Dike, 2003; Pulos, 1978).

Wikinfo, an internet encyclopedia defined at a broader level that industrial design is an applied art which considered aesthetic, usability of designed product as paramount with such details as colour, texture and ergonomics. This is viewed from choice of materials presentation to final consumers, that is, it has a focus on concepts, products and process.

In the United Kingdom, "industrial design" implies design with considerable engineering and technology awareness alongside human factors which is a "total design approach".

Based on the brief definitions above, it is obvious that industrial design embraces the production of prototypes with adequate consideration for aesthetic appearance, function and industrial processes.

Prof. S.A. Adetoro informs that the concept of industrial design as a course of study in Nigerian higher institutions was first muted in 1977 when it was strongly felt that products from the applied arts programmes lacked the knowledge of mass production techniques. He was of the view that the concept of industrial design that was based on the peoples' culture would be more appropriate. The new Universities of Technology which emerged subsequently adopted the nomenclature and tailored their curriculum towards meeting the expectations in the various industries.

An attempt was made to classify industrial design practice in Nigeria; to be situated under the applied art programme which is craft based (Ogunduyile, 2007). However, it was noted that applied arts as being practiced in Nigeria is devoid of industrial processes and necessary facilities. He further observed that the concept and roles of industrial design were not understood for four decades after its introduction into the Nigerian educational system. Therefore, this lack of understanding has impacted negatively on the role of industrial design in industry, business, economic planning strategies and global market place which other nations have taken advantage of.

The carving out of Industrial Design from Department of Fine Arts, Ahmadu Bello University in 1977 was in an attempt to respond to the work demand and challenges from university graduates whose job opportunities were increasingly becoming narrowed down as a result of intense social and economic change being witness at that time (Akinbogun, 2004; Ogunduyile & Akinbogun, 2006). There was a sudden boom experienced from oil revenue in Nigeria and this led to the emergence of new industries that needed to create fresh and creative products. The increase in population which requires faster methods of production in graphic outfits, textiles and ceramics industries encouraged the carving out of the programme. Considering the lofty idea of the programme, more industrial design institutions had since been established. These include the Federal Universities of Technology located at Yola, Bauchi, Akure and Ogbomosho. The National Universities Commission provided the baseline curriculum which dwelt among others on the:

a. provision of professional education to designers who could solve complex industrial problems;
b. Development of students understanding and awareness of the social, cultural, physical, technical and economic activities of the Nigerian society;
c. Development of students ability to provide appropriate solutions to technological, economic and aesthetic solutions of the Nigerian society; and
d. Involvement of universities in the process of exploiting designs and production problems in national industries generally. These universities were given the mandate to review their curricula in line with National University Commissions (NUC) from time to time.

There have been no clear lines of demarcation between the fine arts courses and that of industrial design. Fine art products are often identified for their decorative and sensuous values. However, when the aesthetics values that are inherent in artistic objects are extended to handcrafted utilitarian materials such as textiles, pottery, metal and jewelry, they are classified as applied art. These arts draw their inspirations from various social, religious and cultural settings of the people (Babalola, 1994). If these objects must undergo a process of regular mass production under a quality controlled manufacturing or industrial process, it is referred to as industrial design. All of these can be in engineering, electronics, woodwork, ceramics, graphics and textile designs.

The graduate of industrial design education in Nigeria had contributed immensely to national development in the areas of employment generation, teaching and research, and industrial development. Since government cannot generate employment for every youth, the practical skill acquired through the industrial design training had enabled the youth to create self-employment in pottery entrepreneurship, photography, handcrafted textiles, printmaking, printing technology, and importantly film production as demonstrated by Nigerian *Nollywood* where industrial designers assist at creating costume make-up, special effects and animation. The graphic artists have contributed to book illustration with indigenous theme that embodies values from local cultural contents. The textiles designers also strives to inculcate culturally inspired motifs into locally produced fabrics, which break the past monopoly of foreign designs adopted from India and Europe.

2.1 Historical perspective

The first attempt to introduce fine arts into the Nigerian educational curricula was in 1897 at the Hope Waddel Training Institute, Calabar (Wangboje, 1969). This institution was established by Free Church Mission in 1895. Before then, Aina Onabolu (1882-1963), a man considered to be the father of Nigerian art, had set the foundation for modern art training in the early 20th century. Aina Onabolu, a London and Paris trained artist, noted that the black people had great potentials and ability to express themselves freely in drawing and paintings. Based on this conviction that art could reach its peak in Nigeria if properly handled, persuaded the colonial government to appoint Kenneth Murray in 1902 to further assist the country to strengthen art training and practice. He advocated that art training should be based on African culture and 'not art for art sake' as was the practice in Europe.

Formal institutionalized training in art started at Yaba Technical Institute later referred to as Yaba College of Technology in 1952. Art as a course of study also started at Nigerian College of Arts, Science and Technology (NCAST) Ibadan, the same year but after two years the programme was transferred to Nigeria College of Arts, Science and Technology, Zaria, Kaduna State now Ahmadu Bello University, Zaria. The formal visual art teaching complemented the existing local art and craft centres such as that of Father Kelvin Carols in Oye Ekiti in Ondo State in the 1940s where extensive experiments in weaving, leatherworks, bead making and wood carving with indigenous craftsmanship took place. All of these works were utilized by Catholic Missions in Yoruba land. This avenue provided adequate opportunities for the training of many local artists such as Lamidi Fakeye, a prominent wood carver in Nigeria.

Ulli Beier inspired the commencement of the Oshogbo Art School in 1963. The main aim was to develop untrained artists. The center's major focus was training artists that could work on local cultures, folklores and narratives with production that fuses cultural traditions with modern practices. Prominent participants in the art school were Jimoh Buraimoh, Taiwo Olaniyi, Mariana Oyelami, Asiru Olatunde and the prominent textile artist Nike Davies Okundaiye. Nike Okundaiye worked under Susanne Wenger at the Oshogbo Art School where Yoruba folklores in form of dreams and nightmares were fully explored.

Formal art institutions were growing in leaps and bonds. The Zaria Students Art Society was formed between 1957 and 1961. The group was made up of creative students who formulated a principle for themselves in resuscitating best of Nigerians traditional culture and harmonizing it with best practices in the world. The following members personified this *Zarianism* spirit: Yusuf Grillo, Simeon Olaosebikan, Uche Okeke, Bruce Onakbrakpeya, Demas Nwoko, Osiloka Osadebe, Okechukwu Odita, Felix Ekeada, Ogbonnaya Nwagbara and Ikpowosa Omogie. The Ori Olokun Cultural center was established at University of Ife, now Obafemi Awolowo University in 1968 with emphasis on Yoruba folklore, the river goddess and spiritual analysis.

The foundation of art training and practices was laid in the 1900s. A substantial development and growth were recorded between 1900 and 1938 when a number of artists with traditional art background began to make their impact felt in the formal setting. Between 1930 and 1970 was a period that witnessed a type of unprecedented transformation and quest for identity and promotion by the emerging artists. Formal art institutions during this period began to produce individuals and art movements which sought to embrace traditional themes into their works. The exploration and fusion of traditional values into their artworks became paramount. It was inferred that Ben Enwonwu and Bruce

Onabrakpeya were among the disciples who encouraged and inspired the younger ones in this direction (Oloidi, 1995). Bruce was able to embrace the use of indigenous design in his printmaking. He established a workshop in Agbara Ottor where young talents are trained on regular basis in material exploration. Between 1900 and 1977, all aspects of visual arts were classified under fine and applied arts. The fine arts were considered mostly from their decoration potentials while the applied phase catered for craft aspects which were often handmade.

3. Conceptual clarification of Cultural Inspired Design Education (CIDE) in Nigeria

Assembly of Alaska Native Educators (1998) defines a Culture-based education as an education which reflects, validates and promotes the values, world views, and languages of the community's cultures. Culture may be defined as people's tradition, history, values and language that make up the culture of a group and which contribute to their identity. Culturally inspired design education can be said to be an education that honour all forms of knowledge, ways of knowing and world views equally.

Culturally inspired designs are expressed in Nigerian context from deep conceptualization of subject matter as well as its ultimate functions. The conceptualization embraces the spiritual and social characters the design accommodates. They are usually embodied in different media with elaborate expressions of designs that are associated with where they are meant to serve culturally. This is exemplified in wood carvings of pillar post made for the palaces, shrines or town halls which depict folklores and mythological concepts.

Culturally inspired design education in Nigeria involves the formation and generation of indigenous patterns and motifs in the design of art works and utilitarian objects. The foci of the industrial design curricula in general education and training of design students and apprentices centered on the actualization of "local content" in all creative materials. Apart from providing professional education to designers who could solve industrial problems, students and designers are involved in creating a highly culturally rooted motifs and symbols that are eventually translated to prototypes and objects. For instance the cultural symbols and motifs synonymous with Hausas of Northern Nigeria are at sharp variance with those found on objects produced by artists in the Southwestern and Eastern Nigeria. A recurring cultural symbol often found in the artistic works of the Northern Nigeria consist of the "Northern knot" motifs (crisscrossed elliptical knots) which signify the bonds of political unity envisioned for Northern Nigeria (Fig. 1), while the Yoruba motifs of Southwestern Nigeria reflect traditional beliefs situated in cultural festivals. The Northern knot symbolizes unity in diversity and is elaborately expressed in the palace art of Northern Nigeria. This have been used as design element in Nigerian currency and as crest on edifices jointly owned by 19 Northern States. The royal palaces in Nigeria are usually attributed with cultural art and symbols which form an important aspect of communicating power and royal splendor. The use of masks, beads, fly whisks and other paraphernalia of worship systems are indices of creative motives of the artists.

It cannot be gainsaid then that the inculcation of the knowledge of the culturally inspired design processes described above in the training of industrial design students is another requirement in the Nigerian curriculum of industrial design programmes. It is to be noted as well that the application of the knowledge of culturally based designs on industrial and mass produced objects and materials such as fabrics, ceramic wares, graphic works, interior decoration and metal smiting, encourage patronage from local and international consumers.

Fig. 1. "The Northern Knot". A symbol of unity in diversity among the Northern States of Nigeria

Fig. 2. An Ancient Ife figurine bronze head. An index of the cultural heritage of the Yoruba people in Southwestern Nigeria. Source: http://www.africanart.org

3.1 National questions on Culturally Inspired Design Education

Nigeria has a rich cultural heritage which was enhanced by its arts and crafts culture. Traditionally, the art and crafts as practiced by the people has been providing the foundation for technological growth. The Nigerian craftsmen provided the various functional and aesthetically pleasing implements in household items, furniture, metal working, farm tools, brass casting, leather works, textiles and a lot more. Design is seen to have an important role in upgrading and development of craft products.

With the increase in population, more exposures to foreign influences, education and products from manufacturing industries, the production from the craftsmen could no longer meet with demands of the people. The concept of mass production brought by industrialization was more than what the craftsmen could comprehend as they now found it difficult to compete with cheap factory products.

Culturally inspired education in Nigeria has engendered the training and education which provide adequate attention to the growth of indigenous technology. The craftsmen considered the culture of the people in terms of concept, forms, motifs, shapes and the local methods required in the production of goods and services.

The introduction of industrial design education in Nigeria was to make easy the application of design methodologies and techniques on craft products which needed to be taken beyond the borders of Nigeria. It was conceived as a program that could move the production processes in craft to a higher pedestal in terms of finess and mass production without losing significantly its Nigerian identity.

The training of industrial designers in Nigeria is often based on a foundation of fine arts and crafts and individuals in the field are expected to be well grounded in basic craft processes, high quality execution and decoration of products.

The role of industrial design in Nigeria could be better appreciated when one considers the persistent high level of youth unemployment. Those who have gone through the program have been considered for appointments in various segments of the economy and have been contributing to national development. The program has made it possible for youths to acquire skills in all aspects of art, crafts and design. With the knowledge acquired, many became self-reliant by setting up design and production studios to provide goods and services for the communities in textiles, ceramics, graphics, jewelries, etc. Industrial design has been responding to local needs using local materials, strengthening cultural identity, fostering market access and providing the foundation needed for productive jobs in small and medium scale production.

An analysis of how much impact industrial design and designers is perhaps very significant, though it was claimed that such impact have been circumvented by inability to consolidate various efforts (Ogunduyile, 2007). In 1997, a national conference was organized by the Department of Industrial Design, Federal University of Technology Akure. It was well attended by artists, designers, engineers and architects who presented many thought provoking papers which gave birth to the first indigenous Nigerian Journal of Industrial Design and Technology (JINDEST).

The conference created an impetus for subsequent emergence of a group of designers in the year 2000 to convene a meeting at the Department of Creative Arts, University of Port-Harcourt, Nigeria. The conference generated a proceeding in form of a book titled *Design History in Nigeria,* edited by J. T. Agberia and was sponsored by the National Gallery of Art, Nigeria. It was highlighted that the conference created the expected awareness and was on record that the developmental trends of design principles and practice were put in their historical perspectives (Agberia, 2002). This gathering also led to the formation of a national body which could probably develop the strength and muster the political will to regulate the practice of industrial design in Nigeria particularly and in Africa generally.

Another meeting was also convened in Ibadan at the Demas Nwoko Center to further create awareness on industrial design at the instance of one of the Nigerian foremost designers and architects in the country in persons of Demas Nwoko and John Agberia. The meeting gave birth to the formation of Association of African Industrial Designers, which was expected to give leadership and direction to design education. The first inaugural meeting of the association was held in Benin in 2001 with the aim of providing opportunity for articulating the place of industrial design in Nigeria and the entire African continent while working towards generating awareness within the discipline. Nwoko apart from breeding new ground in theater design and season of wood including its use for furniture and interior design, the quintessential qualities of his building at Ibadan combined earth (mud) with saw

dust which ultimately became recognized and accepted by architects including governments.
The forth national conference was also convened by the Culture and Creative Art Forum (CCAF) in November 2006 at the Federal University of Technology Akure to address the role of design in a dynamic society. The confab discussed the role of Art, Design and Technology in the 21st century with such sub-themes as: *Artists and the Challenges of Industrial Technology; Arts and Design* and *Nigerian Environment* and *Art and Design as Creative Enterprise*. The objective of CCAF that year was to provide opportunity for critical discourse on matters and issues that borders on development of Art, Design, Culture and Technology especially in Africa.
The formation of art movements which advanced various forms of art concepts devoid of imperial influence since independence have significantly repositioned the contemporary efforts of younger generation of designers. The Culturally Inspired Design Education (CIDE) effort engendered by cultural concepts with sustainable development initiative for social and economic revival is a welcome development in line with the advocacy of the New Partnership for African Development (NEPAD).

4. The new paradigm

This would be considered in term of contributions made so far by Culturally Inspired Industrial Design Education to national development through job creations, teaching at both formal and informal levels, highlighting various research efforts and individual contributions.
There are constant urge to create products and projects that enables self-expression, emotional connections and a more sustainable designs that have cultural meanings and encourage business patronage model which continue to attract enviable client list .Works were produced in highly innovative areas that includes pottery, textiles, printmaking, graphics, fashion design and photography. The Culturally Inspired Industrial Design Education in Nigeria has been adapted as a major agent of change. A few works of students and professionals are hereby reviewed.
Of interest in Nigeria are the ceramic cottage industry in Ushafa, Bwari and Dajo. The first two are situated at the Federal Capital Territory, Abuja while Dajo is located in Makurdi in Benue state. Ushafa and Bwari have Saidu and Stephen Myhr respectively as their chief potters. The two centers engage women potters who are good in the traditional method of production who made very relevant contribution by adopting contemporary method using rice husk ashes, granites, wood ashes, kaolin, and feldspar etcetera as their source of ceramic glazes which are applied on ceramics wares. Of significant impact is the knowledge passed to students from various tertiary institutions in Nigeria during Industrial Work Experience (SIWES) by the cottage ceramic industry. The opportunity gives the students the ability to mass produce and also give finishing touches using indigenous symbols and motifs as used by the local potters. Dajo pottery in Markurdi is known for modern production of ceramic pieces.
Extensive teaching and research work had been carried out on intaglio printmaking at the Federal University of Technology Akure, Nigeria (Etsename 2007). Some of the prints that were made with students are mass produced with such themes as the *African child*; the *Fulani Nomadic life;* and some other title that depicts special messages borrowed from everyday life and traditional folklores (Fig. 3, 4, & 5). Fig 3 '*the drifters*' visualizes a scenario

that describes a major phase in the Fulani nomad's life of Nigeria. The seasonal experience during the harmattan period (dry season) of Northern Nigeria, forces the nomads to migrate to the Southern part of the country where they find pasture for their herds. This transient nature has created an avenue for a wide view of his world which has been influential to their way of life. Figure 4 *Mai Nono 1* (the milk maid) is a scene that depicts the plurality of the Fulani woman's personality in playing economic, social and domestic roles. *Mai Nono 2* in Figure 5 depicts an elderly Fulani women engaged in the hawking of dairy products derived from their herds. All these works are the printer's medium of communication towards enlightening the Fulani tribes of Nigeria in relations to improvements in their socio-educational, economic and domestic life.

Printmaking using wood as media was also explored by Oladumiye Bankole, a Graphic Designer and educator (Fig. 6-7). Figure 6 portrays the role of the king's trumpeter at heralding information within a kingdom while figure 7 shows a vital role of decision makers "the Elders" in the traditional society. These works emphasize the significance of communication and synergetic role for the running of a socio-political society. Other research effort had been vested into use of graphics design in establishing relationship between the use of advertisement and consumers urge to purchase products in Lagos State (Kayode, 2010). It was established that languages of local expression used on outdoor billboard affects the urge to purchase products. The use of native language to target audience on outdoor bill board campaign in Lagos State was also advocated for. Ogunduyile Sunday, a textile designer and design educator has worked on a number of textile projects which incorporated African motifs and symbols. (Fig. 8) "Opon Ibile" shows traditional panel consisting motifs derived from cultural items used by the people, while the "Osupa" (Fig. 7) depicts the concept of the moon. Chief Monica Nike Okundaiye is a celebrated textile artist who is in love with "indigo", a traditional material used for dyeing fabrics. She is based in Osogbo, Osun state, Nigeria and have trained many students and apprentices both in Nigeria and abroad especially in Italy (Fig. 10). Her mentoring and art classes offered hope and new livelihood for many young ladies and women. Her lessons in artistic enterprise are wonderful illustrations about how Nigerian creative industry is making positive impact on the country. There are also recent interface of design education in ceramics with science, technology, and engineering such as Exploitation and Adaptation of Bio Gas to Ceramic Kiln by Yusuf Sadiq Otaru; study of the Qualities Of Alkaleri Kaolin In Fired Ceramics by A. D. Umar; Utilization Of Local Raw Materials for the Reproduction of Dense Alumino-Silicate Refractive Bricks for Furnace Using Semi-Dry Processing Techniques by Umar Sullayman and Production of High Fired Porcelain Bodies and Glazes by I. B. Kashim (Umar, 2010). Figures 12 and 13 are ceramic pieces made by O. S. Adelabu and J. O. Ohimai respectively with reflections of cultural symbolism. All of these are reflections of Culturally Inspired Design in Nigeria. Local industrial products in recent times have incorporated culturally inspired designs to replace foreign concepts. Book illustrations with cultural inclinations are done by graphic designers for local publishing outfits like the Academic Press based in Lagos, Nigeria.

Culturally inspired design education goes beyond the adoption of cultural events and traditional skills into the creative and cultural art curriculum in Nigerian schools. The objective of culturally inspired design education is to promote students awareness about their culture. The institutional recognition and validation of its societal culture helps students to be conscious of their cultural endowment and to appreciate the accomplishments of their family, their community and their descendants. This helps in building a sense of dignity and nationalism.

Fig. 3. "The Drifters". PVC Relief Print (Etsename L. E., 2002)

Fig. 4. "Mai Nono 1". PVC Relief Print (Etsename L. E., 2002)

Fig. 5. "Mai Nono 2". An Intaglio Plastographic Print (Etsename L. E., 2002)

Fig. 6. "The King's Trumpeter". Printmaking using wood (Oladumiye B., 1999)

Fig. 7. "Council of Elders". Printmaking using wood (Oladumiye B., 1999)

Fig. 8. "Opon Ibile". Traditional Textile Designs with Batik (Ogunduyile S. R., 2010)

Fig. 9. "Osupa" Traditional Textile Designs with Tie Dye (Ogunduyile S. R., 2010)

Fig. 10. Trainees at Nike Art Centre in Oshogbo, Osun State, Nigeria (2010)

Fig. 11. Traditional handcrafted Fabric Designs at Nike Art Center, Oshogbo, Nigeria (2010)

Fig. 12. A Wheel-Thrown Piece with Hausa Traditional Geometry Design. Bisque 15" high (Adelabu O. S, 2007)

Fig. 13. Hand-built ceramic mask. Glazed 12" high (Ohimai J. O., 2008)

5. The challenges of Cultural Inspired Design Education in Nigeria

Craft in the context of globalization represent a balance between preservation of tradition and global awareness of the diversities of culture. The traditional handcraft in Nigeria is one of the sustainable options for producing objects and artifacts using natural and local materials. The introduction of formal training in industrial design has complemented the role of traditional craft production culture. The problems facing the development of culturally inspired design in Nigeria is classified under the following:

- The negative effects of colonialism when the indigenous craftsmen who would have provided the foundation for technology were relegated to the background. It is widely believed that the emergence of colonialism affected negatively the development of traditional creativity and craftsmanship in favour of Western culture. Western education was criticized for establishing schools with curricula that could not advance the course of traditional crafts and technology.
- The negative attitudes of Nigerian society towards the choice of arts and design as a course of study and the preferential treatments given to those in science and technology by government over fine art. This problem of psyche have led to shortage of specialized teachers in schools to teach cultural and creative art subjects which are considered by many nations as one of the viable subjects capable of reducing poverty and bringing about sustainability of social and economic development.
- Inability of government to introduce appropriate design policies because of lack of understanding of the advantage of the programme. As a result the negative attitudes, appropriate organs of government were not put in place to articulate the manpower needs of the various industries. Hence, the setting of industrial design objectives did not take into consideration the views of experts and stake holders in the field.
- Non commencement of viable art and design programme at the primary and secondary school levels. The inability to make the field attractive to teachers and learners did not favour the harnessing of a holistic creative endeavour necessary for cultural and entrepreneurial development. As it is, the role of art has been too overemphasized in our educational system as a medium of unity and instruction for all subjects among others, but it has not being given adequate time and attention for proper training of the subject matter.
- Synthesis of literature reviews indicate that many schools have no art subjects in their curricula and art teachers are not even available where it is taught (Ubangida2004; Barnabas 2005).
- The relevance of art and design curricula to the world of work, attitudes of the business community and private organizations to industrial design are not in consonance with the expected roles of industrial design in the context of African value and culture.
- There is lack of competent teachers who are grounded in art and technical disciplines to propagate the significance of the development of cognitive, affective and psychomotor skills of the students through a culturally inspired education.
- Design programmes in schools are under-funded and as such the advantages inherent in the programme are undermined at the expense of its usefulness to the students, economy and government. (Ogunduyile, 2007)

6. Suggestions for improvement on Culturally Inspired Design Education

In order to achieve the objectives of Culturally Inspired Design Education for overall national development and sustainability at all levels of education, government, private concerns and stakeholders should work together. This would enhance the unity and coherence needed to make available the required human and financial resources that could improve the quality of art and design programme. It is therefore recommended that:

- The redefining of goals and scope of industrial design programme will go a long way in matching the talents of its stakeholders and their skill with what obtains globally. The need to redefine design activities in consonance with socio economic context becomes imperative.
- Culturally Inspired Design Education would thrive in the 21st century as a vehicle for self -reliance if art and design programme is reviewed at tertiary level while it is revitalised at the primary and secondary school levels. The objectives of the Nigerian University Commission (NUC) benchmarked minimum academic standard for industrial design programme should be well articulated and implemented to the letter. Art and design education should be seen as a viable tool to advance the course of science and technology. Hence, creative art subjects should be strengthened at various educational levels while there should be adequate provisions for art and design equipment coupled with influx of qualified teachers.
- Culturally Inspired Design Education must encourage crafts sensibilities. This could be fostered by re-introducing crafts and cultural studies into the primary school curriculum in order to rebuild the foundation of knowledge and proper appreciation of the national cultural heritage. A proper orientation about the relevance of art and culture must be advocated for, so as to pave way for development of culturally inspired design solutions and enable people to be culturally sensitive. Those who get such exposures penetrate into many occupations to make a huge difference to their professional lives by showcasing the impacts of their cultural backgrounds.
- There is a need for value reorientation on the role of art and culture in the development of the nation. The effect of civilization which has watered down the core values and significance of culture in the minds of the people should be remedied. Besides, a point of equilibrium must be set between science, technology and art. Therefore, it is expedient to sensitise the public to have a new outlook about the content and context of the culture in order to save it from its current place of relegation and misinterpretation.
- There is the need to involve non-governmental organizations in the propagation of the importance of culturally based design education in Nigeria. There should be formation of design institute that will be strategically positioned to influence government at enacting legislative policies that can harness culture as a tool for national development. Culturally Inspired Design Education would only thrive if serious design policy that could bring value to technological and cultural heritage is brought to the fore by government in line with the objectives of the New Partnership for African development (NEPAD). Such policy must be all embracing, holistic and integrating in terms of sustainable development initiatives for social and economic revival of Africa as championed by prominent Nigerian artists and designers such Demas Nwoko, Bruce Onakbrakpeya, Yusuf Grillo, Jimoh Akolo Nike Okundaiye among host of others. The institute will be responsible to maximize the advantages inherent in Nigerian cultural heritage to be able to forge a common front for culturally inspired design education in all its ramifications.

The Association of Canadian Industrial Designers (ACID) was founded in 1948 to promote and represent the interest of its corporate and professional members to government and other international associations such as the International Council of Society of Industrial Designers (ICSID). The association is dedicated to increasing the knowledge, skill and proficiency of Canadian industrial designers through maintenance of close contact with its corporate members as it represents them on both national and international level. The association also promotes the use and value of industrial design to industry and the public. This aforementioned establishment stands as a formidable model of a synergetic effort that can inspire a new paradigm towards promoting Culturally Inspired Design Education in Nigeria.

7. Conclusion

Culturally Inspired Design Education in Nigeria in contemporary times indicates that culture is a viable element in the study of Industrial Design both at the formal and informal sectors considering its growing contribution to the economic life of the nation. Nigeria has witnessed far reaching changes in the way people live as influenced by technology, improved communication and emergence of a classless society inspired by new ideas. Modernity in processes and products are achievable when design concepts are culturally inspired.

8. References

Adetoro, S. A. (1990). Creation of Industrial Design Department in Nigeria: Ahmadu Bello Experience, In *Creative Dialogue,* D. Jegede (Ed.), Pole-Tobson, Lagos, Nigeria

Ahuman, A. M. (1993). Clay as a Creative Media: A Survey of the KimKim Musical Instrument from South Kaduna, In: *Diversity of Creativity in Nigeria*, B. Campbell, R. I. Ibigbami, P.S.O. Aremu & Agbo Folarin, (Ed.), 143-152, Bols Creation, ISBN 978-32078-0-6

Agberia, J. T. (1996). The Role of Women in the Development of Nigerian Pottery Art: Abatan and Kwali as Paradigm of Two Traditions, pp. 156-163, ISBN 978-2458-39-2

Agberia, J. T. (2002). Design History in Nigeria: Critical Issues of Contemplation, In *Design History of Nigeria,* J. T. Agberia, 1-12, National Gallery of Art Nigeria & Association of African Industrial Designers, ISBN 978-33527-7-2, Abuja

Arlene, B. (2005). Traditional Crafts Create Connections: Adding Gezelligheid to Design in an Era of Globalisation. Dutch Design Publication

Assembly of Alaska Native Educators (1998). Alaska Standards for Culturally Responsive Schools. Fairbanks: Alaska Native Knowledge Network, University of Alaska Fairbanks. Retrieved from http://www.ankn.uaf.edu/publications/culturalstandards.pdf

Babalola, O. (1994). Problems of Perception and Definition of Contemporary Nigerian Art, In African Art-Definition, Forms and Styles, R. Kalilu (Ed.), 53-61, Ladoke Akintola University of Technology Ogbomosho, ISSN 1118-8154

Barnabas, S. D. (2005). Teachers Assessment of Children Creative Art Works. Case Study of Some Primary Schools in Kaduna State. Unpublished M. A. Thesis, Department of Fine Arts, Ahmadu Bello University, Zaria.

Dike, P. C. (2003). Industrial Design Development in Contemporary Africa. A Keynote Address Presented at the Maiden Edition of the Industrial Design Conference, Imo State University, Owerri

Ekong, C. E. & Ekong, I. D. (2009). Industrial Design Culture: An Imperative in Nigeria's Technology Development. Art in Contemporary Nigeria: Its Value and appreciation, NGA Nigeria, pp. 171-177

Egonwa, O. (1995). Patterns and Trends of Stylistic Development in Contemporary Nigerian Art. *Kurio-Africana*, Vol. 1, No. 2, pp. 3-12

Egonwa O. (2001). The Evolution of the Concept of Natural Synthesis. *USO- Nigerian Journal of Arts*, Vol. 3, No.1 & 2, pp. 52-59, ISSN 1117-9993

Etsename, L. E. (2007). A Socio-Cultural Exposition of the Fulani Nomads in Nigeria. *Art Mediterraneo* (An International Journal on African Art), Via Gamberi 4, 400317 Sasso Marconi (BO), Issue 58, pp. 46-54

Irivwuri, G. O. (2010). An Appreciation of the State of Visual Arts in Nigeria (1900-1970). *Anthropologist*, Vol. 12, No. 2, pp. 113-117

Hank, K.; Edward, D. and Belliston, L. (1978). *Design Yourself*, William Kanfman Inc., USA

Kayode, O. F. (2010). Relationship between the Use of Advertisement and Consumers Urge to Purchase Products in Lagos State. A Paper Presented at 1st International Conference, School of Environmental Technology , Held at the Federal University of Technology, Akure, Nigeria, 25th-27th October, 2010

Mamza, P. M. (2007). Contemporary Issues in Fine and Applied Arts in Nigeria. *Multidisciplinary Journal of Research Development*, Vol. 8, No. 4

Oloidi O. (1986). Growth and Development of Art Education in Nigeria (1900-1960).*Trans-African Journal of History*, Vol. 15, pp. 108-126

Ogunduyile, S. R. & Akinbogun, T. L. (2006). *Industrial Design Status and Challenges to National Development in Nigeria,* Retrieved from *https://www.sabs.co.za/content/uploads/files/Nigeria.pdf*

Ogunduyile, S. R. (2007). Industrial Design Concept in Nigeria and the Challenges to National Development. A Paper presented at the Network of Industrial Design Meeting, South Africa

Ogunduyile, S. R. (2007). Challenges of Curriculum Development and Implementation in Industrial Design: A Case Study of Nigeria and South Africa, In: *Rethinking the Humanities in Africa*, B. Akinrinade, D. Fashina, D. Ogungbile, J. O. Famakinwa (Ed.), 221-237, Obafemi Awolowo University, ISBN 978-978-080-221-7, Ile-Ife, Nigeria

Oloidi, O. (1995). Art and Naturalism in Colonial Nigeria, In *Seven Stories about Modern Art in Africa*, H. Jane (Ed.), 193, White Chapel Gallery, London

Ojo, E. B. (1991). Some Aspect of Yoruba Myths and Cosmogonies in Contemporary Printmaking. Unpublished MFA Thesis, University of Benin, pp. 80.

Pulos, A. J. (1978). *Opportunities in Industrial Design*. VGM Career Horizons, a Division of National Textbook Company, Stokie, Illinois, U.S.A.

Sheba, E. (1993). Art and Creativity: A Quest for Meaning, In: *Diversity of Creativity in Nigeria*, B. Campbell, R. I. Ibigbami, P.S.O. Aremu & Agbo Folarin, (Ed.), 167-172, Bols Creation, ISBN 978-32078-0-6

Umar, A. (2010). Thirty Years of Industrial Design in Nigeria. A Paper presented on the occasion of 3 Decades Anniversary Celebration of Industrial Design: Conference, Exhibition and Home Coming Events held at the Department of Industrial Design, Ahmadu Bello University, Zaria

Ubangida, M. B. (2004). An Evaluation of Art Programmes in Some Selected Secondary Schools in Taraba States, Nigeria. Unpublished M. A. Thesis, Department of Fine Arts, Ahmadu Bello University, Zaria

Wangboje, S. (1969). Some Issues on Art Education in Africa: the Nigerian Experience. INSEA-Education through Art and Humanism in a Technological Age, pp. 74-84.

William B. (1973). *African Art in Cultural Perspective: An Introduction*, W.W. Norton, New York, pp. 3

Part 3

Innovation in the Design Process

6

Biologically Inspired Design: Methods and Validation

Carlos A. M. Versos and Denis A. Coelho
Universidade da Beira Interior
Portugal

1. Introduction

Design inspired by nature, bionic design, biomimetism, biomimicry, or biologically inspired design, despite having been a source of inspiration for design activities for a long time, have recently, under pressure from sustainability concerns, gained a role as part of a standard set of approaches to deal with design problems. Nature provides an important model to find solutions to the ecological crisis. The aim of this chapter is to establish a comparison among a set of design methods, meant to guide industrial designers in carrying out activities leading to bio-inspired design. The results of literature review are presented, with emphasis drawn on existing documented approaches to design inspired by nature, and the presentation of the methods, on which a comparative analysis is established. The parameters for the comparative analysis are set out, based on five general goals that are considered applicable to design problems, within the realm of industrial design. The presentation and explanation of the comparisons is followed by a discussion on their implications for theory and practice.

The present day's urgency in achieving environmental sustainability has promoted renewed interest on gathering inspiration from nature in order to create novel design concepts. Design endeavours in several technical disciplines may lead to ground-breaking new concepts when natural systems are considered as a source of inspiration. The focus of this chapter is on joining a bio-inspired approach to the creation of industrial design engineering concepts with a systematic approach to design. The conduction of industrial design engineering projects is inherently structured and supported by methods set forth in the systematic design literature (e.g. Hales 1991, Hubka & Eder 1992, Roozenburg & Eekels 1995, Pahl & Beitz 1996, Ulrich & Eppinger 2004). Hence, in order to be useful and of practical value to the generation of industrial design engineering concepts, bio-inspired design methods should be able to fit into design endeavours that follow a systematic approach to design.

The main purpose of bionics is to carry out a benchmark of nature, of what it created, tested and has evolved over millions of years, in order to improve what man creates artificially (Benyus 1997). A number of design methods, intended especially to guide industrial designers in carrying out the development of biologically inspired design, have been proposed. The chapter establishes a comparative analysis between five methods, retrieved from literature. The methods are presented in similar depth, and the parameters of analysis are also described. The five bio-inspired design methods discussed, following an analytical direction that involves seeking inspiration in nature to solve a given problem, were retrieved from literature and are summarily presented. These methods are analysed in this chapter with regard to their perceived capacity to support the satisfaction of the five chosen high level design aims. These aims were selected considering their high degree of perceived relevance to industrial design engineering problems. A critique of the bio-inspired design methods retrieved is laid out, informed by comparison between the methods regarding their ability to support the satisfaction of the goals. The analysis is based on the scrutiny of the five methods, in relation to the support given towards the satisfaction of five goals, considered of paramount importance, and which are present in typical design projects, albeit translated into a number of requirements, specific to the problem at hand. The comparative analysis is intended to support designers in the process of selecting a design method that is adequate to the problem at hand. The analysis also identifies goals where the methods considered offer no or reduced support for their satisfaction, hence identifying the need for novel methodological proposals. The need to integrate validation activities in the bio-inspired design processes is also emphasized as a result of the analysis and followed through by the proposal of explicit procedures for validation of the satisfaction of goals sought by those pursuing biologically inspired design. This approach is intended to enable the evaluation of outcomes attained with the use of bio-inspired design methods, offering methodological support to designers in order to pursue the validation of bio-inspired concepts generated by them. These validation procedures are demonstrated in a specific design case with the purpose of exemplifying the application of the validation steps proposed. The requirements initially considered for the development of the product functionality considered in the case are also presented and a solution that is proposed to fulfil these requirements, generated using a bio-inspired approach, is evaluated, according to the validation approach presented.

The deployment of the validation process proposed is done within an iterative design case, consisting of a novel CD rack, which draws inspiration form nature, as its main solution principle is inspired on the spider-web. The process of validation makes use of surveys, conceptual-analytical arguments and standard engineering design procedures.

2. Bio-inspired design

The term bionic system, or bio-inspired systems, generally has two usual interpretations, concerning different application domains. The popular interpretation, based frequently on science fiction, is associated to more or less fantastic super-powers, to cybernetics and to robotic creations or additions to organisms. In this line of thought, bionics is presented as a science uniting biology and mechanics, producing devices that capacitate human beings with enhanced powers, whether to compensate for innate or acquired physical limitations, or for mere enhancement. Besides this interpretation, the term bionics is associated with the original meaning of biomimetism (bios - life, mimesis - imitation). According to Benyus (1997), biomimetism is a way to see and value nature, representing a novel mindset based not on what can be extracted from the natural world, but what can be learnt from it. This interpretation is the one of concern in this contribution. In this view, the main purpose of bionics is to carry out a benchmark of nature, of what it created, tested and has evolved over millions of years, in order to improve what man creates artificially.

In the following sub-sections, the origins and evolution of bionics are summarily reviewed, while recalling a few well known examples of bio-inspired design solutions. Arguing for the growing importance of design inspired by nature for industrial designers, the section ends with the presentation of five bio-inspired design methods that will come under scrutiny in the remaining sections of the chapter.

2.1 Origins and evolution of bio-inspired design

Although the terminology of this area of design is relatively recent - appearing for the first time in the U.S.A. in 1958, by the hand of Jack E. Steele (Lloyd, 2008) - the practice, creation and inspiration through learning about nature comes from the most remote and pre-historic times. Primitive human beings used bone harpoons, which were serrated on their edges, to improve their piercing ability. This feature was likely inspired by animal teeth.

Leonardo da Vinci was probably the first systematic student of the possibilities of bionics (Lage and Dias, 2003). From the classical times of the Icarus legend, to the drawings of Leonardo da Vinci, man's dream to fly originated in the observation of bird and insect flight. Leonardo da Vinci realized that the human arms were too weak to flap wings for a long time, and hence developed several sketches of machines he called ornitopters (Kindersley, 1995). Human flight would only be possible in the XXth century, with the aid of the internal combustion engine and the propeller, but the inspiration from nature is anyway at its onset.

One of the most disseminated examples of bio-inspired design is Velcro, invented in 1948, by Swiss engineer George de Mestral, from inspiration he got while observing thistles and the way they got caught in his dog's tail and adhered to clothes. In current times, designers Luigi Colani and Ross Lovegrove have been instrumental in portraying the use of bionics in their creations. Colani became notorious by the use of biodynamic forms in products such as automobiles and airplanes, during the second half of the XXth century (Pernodet and Mehly, 2000). Lovegrove's designs typically demonstrate a link between organic shapes and material science (Lovegrove, 2004). While a bio-inspired approach to design may not represent a universal tool that is applicable to any problem, it may provide support to design activities (Colombo, 2007). A set of five bio-inspired approaches to design, documented in literature, are presented in the following sub-section.

2.2 Methods for bio-inspired design

The goal of bio-inspired design methods consists in offering designers an organized process in order to attain a model that may be applied in design, inspired by the relations between form and function in nature (Colombo, 2007). Despite the success attained in several cases from the use of this approach in design, the bio-inspired design approach may still have room for improvement, in order to become more systematic. Five existing methods have been collected from literature and are presented in Tables 1 to 5.

The design method presented in Table 1 emphasizes the importance of environmental and economical sustainability factors in the development and evaluation of the project by the designer. This method shows little support for organization problems. The method presented in Table 2 provides a detailed description of the procedures involved in natural sample collection and analysis. It also prescribes completely listing the working principles of the natural system. However, this method does not include any procedures concerning the design transfer of the features found in the natural samples. The design method presented in Table 3 gives emphasis to the product life cycle, by giving consideration to issues such as manufacturing processes, packaging and recycling of the product in development. In this method, iterations are implicit, and evaluation of the result of every step is also recommended.

Phase	Description
1. Analysis	Choice and analysis of a natural system. The purpose of this phase is to understand the form, structure and functional principles of the natural system.
2. Transformation	Extrapolation of mathematical, geometrical and statistical principles through a process of abstraction and simplification. Transformation, by the analysis of the analogy, of the characteristics of the biological system into technical and mechanical terms.
3. Implementation	Implement the principles of the relationship between form and structure found in the natural system analysis, for the development of new products.
4. Product development	Development and evaluation of a new product taking the environmental and economic factors for all life stages of the product into account.

Table 1. The Aalborg bio-inspired design method (Colombo, 2007).

Phase	Description
1. Identification of need	Identification of an unmet need in a satisfactory manner and that allows the satisfaction of a particular problem and accurately, for subsequent analysis of the environment in search of potential solutions.
2. Selection and sampling	Practical process step involving the selection of samples in nature that fit the problem and the need at hand. Involves the search for samples in nature and some knowledge about the habitat of the samples to be collected and of the equipment to be used for the collection.
3. Observation of the sample	Observation and analysis of the components of the morphological structure, functions and processes, of the distributions in time and space and of the relationship with the environment. Classification of the sample.
4. Analogy of the natural system with the product	Through the information of functional analysis, morphology and structure, the designer has the capacity to start considering the possibility and feasibility of application of an analogy between the sample studied and the product to design.
5. Design implementation	Considering the feasibility of application of the sample characteristics to the design and from the functional, formal and structural analysis, as well as the needs and requirements of the proposed product, an analysis of the system is held at this stage.

Table 2. The biomimicry design method (Junior et al., 2002).

Phase	Description
1. Identify	Development of the Design Brief for a human need with the details and specifications of the problem to be solved.
2. Interpret	Biological view of the problem. Questioning the Design Brief from the perspective of nature. Translation of the functions of the project into tasks performed in nature.
3. Discover	Find the best natural models to answer / address the challenges posed.
4. Abstract	Select the "champions" with the strategies most relevant to a particular challenge of the project.
5. Emulate	Developing ideas and solutions based on natural models to mimic aspects of form, function and of the ecosystem as much as possible.
6. Evaluate	Evaluate the design solution considering the principles of life. Identify ways to improve the design and bring forward questions to explore issues such as those related to packaging, marketing, transportation, new products, additions and refinements.
7. Identify	Develop and refine design briefs based on lessons learned from evaluation of life's principles.

Table 3. The spiral design method (Biomimicry Institute, 2007).

Phase	Description
1. Problem definition	Selection of a problem to solve and performing further definition of it through functional decomposition and optimization.
2. Reframe the problem	Redefining the problem using broadly applicable biological terms. Asking the question: "How do biological solutions perform this function?"
3.Biological solution search	Find solutions that are relevant to the biological problem with techniques such as changing constraints, analysis of natural champions of adaptation, variation within a family of solutions and multi-functionality.
4. Define the biological solution	Identify the structures and surface mechanisms of the biological system related to the recast function.
5. Principle extraction	Extraction of the important principles of the solution in the form of a neutral solution, requiring a description that removes, as much as possible, the various structural and environmental constraints.
6. Principle application	Translation of the bio-inspired solution principle extracted into a new area, involving an interpretation of a domain space (e.g., biology) to another (e.g., mechanics) by introducing new constraints.

Table 4. Bio-inspired design method (Helms et al., 2009).

Phase	Description
1. Biological solution identification	From the observation of natural phenomena on a macro scale and / or a micro level, a potential solution to apply is sought to transfer to a human problem.
2. Define the biological solution	The components or systems involved in the phenomenon in question are identified in order to outline the biological solution in functional notation.
3. Principle extraction	From the analysis of the biological solution in schematic notation, the fundamental principle of the solution is extracted.
4. Reframe the solution	In this case, reframing forces designers to think in terms of how humans might view the usefulness of the biological function being achieved.
5. Problem search	Whereas search in the biological domain includes search through some finite space of documented biological solutions, the search may include defining new problems (this is much different than the solution search step in the problem-driven processes).
6. Problem definition	By analogy with the definition of the solution in schematic notation, the problem is outlined similarly. The aim is thus to establish a parallel between the systems and components of the biological solution and the problem.
7. Principle application	Once the solution principle is established, it is transformed into a working principle of the technological concept that is needed. This activity will culminate in the embodiment of a bio-inspired solution of a technological product or system.

Table 5. Bio-solution in search of a problem method (adapted from Helms et al., 2009).

For the method presented in Table 4, the process of problem definition and searching for biological solutions is supported by elucidative techniques, suggestions and practical examples. The method presented in Table 5 supports an iterative formulation of the bio-inspired design principle.

The application of bionic principles in a design project can be accomplished by following any of two opposing directions: finding a solution to a problem in nature, or looking for a problem for which a solution has been found in nature. The former approach starts with the identification of a problem (human applications, such as developing or improving products or services) or the need of a project, followed by looking for inspiration from nature or an analogy to foster a solution to the problem (a bionic solution proposal). This approach is well suited to designers seeking inspiration for the development of a particular product. The other approach is based on the observation of nature and its structures in order to collect useful information (bionic inspiration based solution) for human applications (design problems to be sought).

3. Generally applicable goals for bio-inspired design

This section presents an analysis of the likelihood of satisfaction of selected goals with the use of the five methods for bio-inspired design, retrieved from literature and presented in the previous section. Five general goals are proposed that are deemed to encompass many of the requirements pertaining to design projects for which inspiration from nature is sought. The goals were selected based on their perceived level of importance and their perceived ubiquitous relevance across design projects, albeit translated into a number of requirements, specific to the problems at hand. Communication effectiveness, form optimization, multiple requirements satisfaction, organization effectiveness and paradigm innovation for improved functional performance are the goals considered.

Effectiveness of communication depends on the sharing of a language that may be based on a code, gestures, or on signal that is appropriate to the activity and context. For effective communication to accrue it is necessary that the message is clearly delivered and received in a timely fashion, without noise, and that it is relevant to the situation or event that is ongoing.

Optimizing the shape of an object or structure can result directly from the balanced satisfaction (with concessions on both sides - trade-offs) of several key requirements, such as the reduction of material and, or, size, or the satisfaction of greater stability, or reduced drag, depending on the targeted objectives. It is not always possible to achieve an optimal configuration, with maximization of all properties due to inherent conflicts that they sometimes impart (e.g. contradiction between low weight and high strength or high volume or stability). Thus, optimization requires that the configuration reached is the one that best addresses the contradictions and conflicts between the desired properties.

Nature is rife with effective solutions in order to enable, in a limited space, a system to perform various tasks or fulfil several functions. Compliance with multiple requirements reflects the achievement of several key points that are inherent to the problem at hand, aiming for viability and profitability of a small number of structures and elements that are to be used in performing more than one function. This simultaneous satisfaction opens the way for consideration of new objectives to add value and profitability to the designed product or system. Compliance with various targets, carried out by a limited set of features, structures or entities implies streamlining for functional efficiency, which will result in resource savings.

The effectiveness of organization depends on the coordination of multiple structures (which also includes communication) for the performance of activities with the need of differentiation. The coordination of multiple entities in joint activity may lead to more effective results than the performance of the activity separately by each entity, such as that "the whole is greater than the sum of its parts". An example of excellent coordination and effectiveness of the resulting organization can be inferred from observation of the natural system comprised of a pack of wolves. The group can hunt animals larger than the wolf, while a lone wolf may only hunt smaller animals or of a scale similar to his. The organization of the roles of each element within the pack is a pre-condition for achieving this result.

Finally, the fifth goal considered consists in achieving change in the conventional paradigm used to implement a feature, replacing it with an innovative paradigm. The latter may be proposed based on the observation of structures, behaviours and, or, processes of nature that enable improved performance of the function or feature. The features can be characterized by transformation of physical state, state association or state hierarchy, to name a few. This goal is deemed to represent one of the most commonly sought goals by designers inclined to use a bionic approach.

3.1 Likelihood of achieving the goals selected by using bio-inspired methods

Considering the five goals presented, the five methods under focus were analysed in terms of their perceived support offered to designers making use of them towards the satisfaction of each goal. In what concerns the effectiveness of organization, a method oriented from the solution to the problem (Aalborg) is considered applicable to support the satisfaction of this goal, demonstrating that there are a few gaps remaining in order to lead to the full satisfaction of this goal. None of the problem-oriented methods analyzed is considered fully adequate to achieve this goal.

With regard to satisfying multiple requirements, methods oriented from the solution to the problem show, from the analysis, gaps in support to achieve this goal. The methods providing guidance in implementing bionic projects in the contrary direction of analysis, are very heterogeneous. While the method of bio-mimicry offers no support for the pursuit of this goal, in the opposite extreme, with considerable support, is the method of bio-inspired design.

When considering the goal of form optimization, one is faced with a relatively homogeneous landscape, with the methods only offering partial support to attain this goal. The exception of the method of spiral design is highlighted, as it is considered significantly applicable in order to achieve this purpose (this method provides guidance in following the direction from problem to solution).

In what concerns the innovation of paradigm for improved functional performance, all analyzed methods provide satisfactory guidelines which can support the achievement of this purpose. This demonstrates that the primary approach that has been recommended for bionic design centres on the functionality. Moreover, except for individual cases, the remaining goals have been given a minor importance. Although the Aalborg and biomimicry method ratings were similar (except for organizational effectiveness), the steps of the latter are more detailed than the ones of the former, and there is a descriptive complementarity between both. The need to integrate validation activities in the bio-inspired design processes is emphasized, as only a few of the methods (the spiral design and Aalborg design methods) entail some evaluation and iteration. The development and testing of improved methods, providing broad support to pursuing a large scope of design goals, with support for validation of the quality of results attained, is hence necessary. The results of the overall analysis are presented in Table 6.

The bio-mimicry design method is only deemed "applicable with shortcomings" with regard to attaining the goals of optimizing form and improving effectiveness of organization. For the first goal, the assessment derives from the absence of iteration in order to pursue optimization (observing the morphological structure is what is suggested in the method that may provide limited support to pursuing this goal). For the second goal, the assessment takes into account that the method supports no direct account of organizational aspects, but only does that indirectly through structural analysis. The evaluation also results in suggesting the applicability of the method to support the pursuance of the goal of paradigm innovation for increased functional performance, and on the other hand, enables suggesting its non-applicability if the goal is to achieve satisfaction of multiple requirements and communication effectiveness.

The spiral design method was granted the rating of "Applicable with shortcomings" with respect to the goal of satisfying multiple requirements. In this method, satisfaction of multiple requirements may take place according to their explanation in the initial specification, if natural models demonstrating the reunion of the functions and, or, qualities sought are analysed. However, the method does not explicitly consider a way to guide the quest to satisfy multiple requirements. The goal of organizational effectiveness receives the same evaluation, as the aspect of organization is not considered directly in this method, but it is only implicit in the consideration of the analysis of ecosystems and natural social conditions. For the other goals at hand, this method proves to be applicable to support their satisfaction if the target is form optimization (especially given the nature of this iterative method, which favours systematic optimization) or innovation paradigm with regard to performance features. No support is perceived to attain the goal of communication effectiveness.

Bio-inspired design methods	Goals sought				
	Communication effectiveness	Form optimization	Multiple requirements satisfaction	Organization effectiveness	Paradigm innovation for improved functional performance
Bio-mimicry (Junior et al., 2002)	Not Applicable	Applicable with shortcomings	Not applicable	Applicable with shortcomings	Applicable
Spiral design (Biomimicry Inst., 2007)	Not Applicable	Applicable	Applicable with shortcomings	Applicable with shortcomings	Applicable
Bio-inspired design (Helms et al., 2009)	Not Applicable	Applicable with shortcomings	Applicable	Not applicable	Applicable
Aalborg (Colombo, 2007)	Not Applicable	Applicable with shortcomings	Applicable with shortcomings	Applicable	Applicable
Bio-solution seeks problem (Helms et al., 2009)	Not Applicable	Applicable with shortcomings	Applicable with shortcomings	Applicable	Not applicable

Table 6. Analysis of perceived support provided by the five bio-inspired design methods selected in attaining five fundamental design goals.

The bio-inspired design method shows gaps in the support offered to designers if the goal is to achieve optimal form, since the focus in this method is set on function. In some design processes supported by the procedures inherent to this method, the search for a biologically inspired solution to perform a given function could lead to considerations of form. However, the method does not provide procedures for optimization and does not explicitly consider form, or shape. The method is also deemed applicable in a satisfactory manner to problems where the targeted goal is either paradigm innovation for improved functional performance, or to satisfy multiple requirements, or a combination of both. However, it is not applicable to support the pursuance of the goals of effectiveness of either communication or organization.

In the Aalborg method, which provides guidance in the direction from the solution to the problem, the degree of applicability to the goals of form optimization and satisfaction of multiple conditions, was assigned as "Applicable with shortcomings". For the first goal, despite the focus on form, there is no effort to optimize. Secondly, because shape, structure and functional principles are considered in this method the, implementation of multiple principles of form and structure may result from the analysis but is not explicitly considered. For the goals of innovation of paradigm for improved performance of functions and for effectiveness of organization, this method is deemed applicable.
For the bio-solution in search of a problem method, which is directed from the solution to the problem, as this method focuses on extracting and implementing the solution principle form nature, both the aspects of optimizing the shape and satisfying multiple requirements, are bound to be sidelined at the expense of the functional principle. The evaluation of this method and the previous one only differ significantly on the applicability to provide support to the pursuance of the goal of organizational effectiveness, because in this method there is no focus on the organizational structure of the biological system centred upon. The method is deemed applicable to support attaining the goal of paradigm innovation for increased functional performance.

4. Validation of goal satisfaction in the bio-inspired design process

The analysis presented in the previous section identified goals where the methods considered were deemed to either offer no support, or only offer reduced support, towards their pursuance and satisfaction. Moreover, only two of the methods (the spiral design method and the Aalborg method) entail some evaluation procedures, albeit limited in scope. This leads to suggest the integration of validation activities in bionic design processes, in order to ascertain whether the desired goals might be met by the use of the concepts generated with the support of bionic design methods. The current section presents a proposed validation approach, summarily depicted on Table 7, based on considering specific validation procedures matching each of the five goals focused in the previous section.

4.1 Bio-inspired design case requirements

Five methods that are intended to support the generation of bio-inspired design concepts, retrieved from literature, were analysed in this chapter. Three of these methods shared a common direction of analysis, which departs from a given problem and seeks the proposal of solutions by gathering insight and inspiration form natural systems. This approach begins with the identification of a problem or the needs of a project, which is followed by looking for inspiration from nature or seeking an analogy with a natural solution to foster the emergence of a solution to the problem (a bionic solution proposal).
A bionic design project was carried out, following an approach combining three of the methods reviewed (Junior et al. 2002, Biomimicry Inst. 2007, Helms et al. 2009). The problem considered was the storage and the physical display to enable browsing of personal music collections, focusing on CDs and DVDs. The conduction of the design process led to seek inspiration form nature, having selected the spider web as a natural example that was the basis for the analogy of working principle established. The requirements established for the project and their corresponding goals are listed in Table 8. Moreover, environmental

Goal	Validation procedures to evaluate goal accomplishment
Communication effectiveness	Validation is made according to the level of communication involved. Passive communication (triggered by observation) - the effectiveness may be evaluated by assessing the degree of the overlap between the meaning intended to be incorporated into the product or system by the designer and the readings of signification made by users or observers (empirical verification). Active communication (synchronous process between a sender and a receiver) - effectiveness evaluated from the assessment of the overlap between the messages from the sender and what is perceived by the receiver, which should conform to what is desired by the sender (empirical verification).
Form optimization	Validation based on a comparative approach with regard to a conventional product with functionality that is similar to the one intended for the bionic concept. Examples: Reducing material and weight - analysis from solid modeling. Stability - static analysis of mass centre (force vector modeling). Resistance for maximum capacity - finite element method and prototype testing. Object storage - capacity, maximum capacity; quantification.
Multiple requirements satisfaction	Validation based on objectively verifying, as much as possible, the level that has been reached for each property implicit in every requirement. This is followed by checking if the resolution of conflicts between non-compatible properties was made with compromises established on every side of the requirements concerned.
Organization effectiveness	Validation based on the comparison between two or more systems performing the same function (including the proposed system), but with different methods of organization. Collect measures of the levels of operation effectiveness from the (real or simulated) systems (including the proposed system), such as execution time, energy expended, material resources expended, or resources generated.
Paradigm innovation for improved functional performance	The evidence of paradigm change depends on the type of paradigm involved. Consider these examples of two kinds of paradigm change: Paradigm change at the organizational level - could involve changing from a centralized model of decision making to a process of cooperative decision making distributed and performed by multiple system elements. Paradigm change at the technical level – could involve fundamental changes in working principle, shape archetype, drive technology or kind of energy supplied. The verification of the satisfaction of this goal may centre on a conceptual-analytical argument distinguishing between the existing and the new paradigm, possibly illustrated by descriptive imagery and, or, technical schemes.

Table 7. Specific procedures suggested for consideration of the processes of validation of goals sought in bio-inspired design endeavours.

concerns were expressed in terms of reduced environmental impact of materials, ease of maintenance and repair, as well as low weight of the product (and its package) for transportation. These requirements were dealt with in the design project, impinging on the selection of materials (selection of a bio-polymer and an organic elastomer) and on the design of the project (impinging on the goal of paradigm innovation for improved functional performance).

Requirements	Goals sought*
1. Nice and appealing shape, enabling the user to develop an aesthetic interest in product	Communication effectiveness
2. Sending a message of an avant-garde character, creative and youthful	
3. Enhanced stability against a dynamic disturbance compared with a conventional solution†	Form optimization
4. Increased lightness compared to conventional solution†	
5. Proper positioning of the title of the CDs, DVDs and books for enhanced readability	
6. Storage with versatility of CDs, DVDs or books	Organization effectiveness
7. Enhanced gripping of objects compared with a conventional solution	Paradigm innovation for improved functional performance

*- Satisfaction of multiple requirements is implicit in the consideration of the several requirements; † - Conflicting requirements, requiring a trade-off.

Table 8. Listing of requirements set for the bio-inspired design case presented and their corresponding goals that were sought.

The validation processes carried out within the exemplified design case used are summarily described, and an overview of the evidence used and the results obtained is provided in the following sub-sections, considering the goals depicted in Table 7. In what concerns the goal of satisfaction of multiple requirements, a conflict was detected between the requirement of enhanced stability and lightness. This conflict was solved by means of an approach akin to TRIZ (Altshuller 1994), with the contradiction solved by change of state, in the second iteration of the design. Bionic tower 2 hence encompasses a reservoir in the basis which may be filled with water or sand for added stability, while lightness is still guaranteed, for the sake of environmental concerns, especially focusing on the production and distribution phases of the product's life-cycle.

4.2 Validation of communication effectiveness

The perception by the user of pleasantness and appeal, enabling the development of an aesthetic interest in the product (first requirement in Table 8) was validated through a questionnaire where, among other things, each of the two bionic CD towers was visually compared, with a conventional tower (Figure 1). The validation of this requirement is necessarily subjective, because the key issue that arises relates to the taste and sensitivity of each individual questioned. Respondents, answering by email, accounted to 85, aged between 18 and 60, both male and female, and with diverse professional and knowledge specialities.

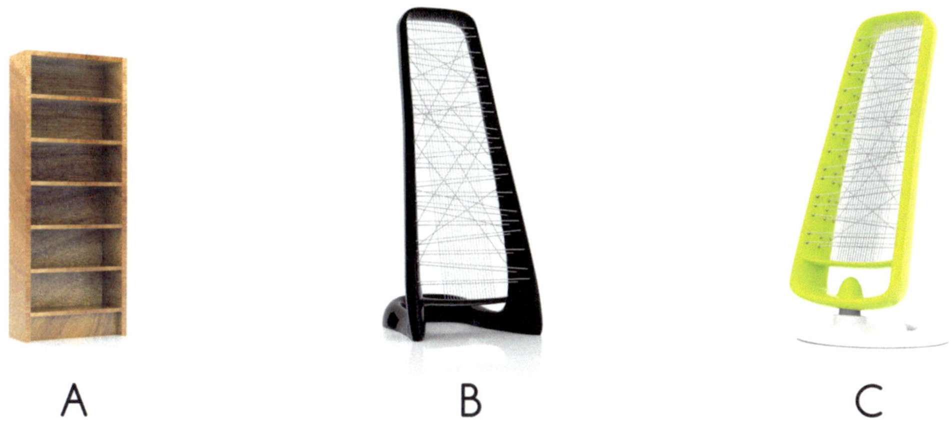

Fig. 1. Depiction of a conventional CD tower, and the two bio-inspired CD tower racks designed: conventional tower(A), bionic tower 1 (B) and bionic tower 2 (C).

Each respondent indicated which of the CD racks was personally more aesthetically pleasing and appealing, from 3 paired comparisons presented. The paired comparisons approach applied to this case of three objects enables 8 possibilities of response, two of which are incongruent, since no ranking of preference can be established out of them. Three out of the 85 respondents reported incongruent paired comparisons. Thus, the analysis of results was carried out for 82 responses. The results were analysed on the basis of the procedure for calculating the Kendall coefficient of concordance (Siegel & Castellan 1988). The average ranking obtained was bionic tower 1 (first place), bionic tower 2 (second place) and conventional tower (third place). This result is considered significant to represent the overall opinion of respondents to a confidence level of 99%. These results support the validation of the first requirement depicted in Table 8. Both the first and second bionic towers received the preference of respondents over the conventional tower, which supports the validation of the gains in terms of pleasantness and aesthetic appeal, for both versions of the bio-inspired design.

In what concerns the second requirement that contributes to the goal of effective communication, validation was sought by means of a technique of anthropomorphizing products through the attribution of personality dimensions. In a first phase, a translation of the requirement into a product personality profile (Jordan 2002) was proposed. In the second phase of the process, a sample of specialized public (eight undergraduate Industrial Design students) assessed the personality profile of the three objects shown in Figure 1. In such, whether or not the message intended by the designer was transmitted to the public could be verified.

The second requirement set in Table 8, was decomposed in a number of concepts to promote the matching process envisaged. This led to considering the attributes of modern, elegant, youthful, joyful, flexible and dynamic. Moreover the attributes consisting of lightweight and stable were also considered from the third and fourth requirements. The correspondence between product attributes intended by the designer to be perceived by the public and

product personality dimensions (Jordan 2002) are shown in Table 9. The outcome of analysis on the respondents assessment of the personality profiles is also shown, based on evaluation of Kendall's coefficient of concordance (Siegel & Castellan 1988).
For every personality pair, analysis was performed as exemplified for the pair energetic - non energetic energy, the average ranking of the panel of respondents (with a significance of 99%, given by the assessment of Kendall 's coefficient) resulted in the following rank order: 1st C, 2nd B, 3rd A. As a conclusion to this result, it is understandable that tower C (bionic tower 2) is considered more energetic than the tower B (bionic tower 1), and that tower C (conventional tower) is considered less energetic than tower B. This means that tower C is deemed the least energetic of the three towers and that C is the tower that emerges as the most dynamic and the less dynamic, thus validating this communication requirement.

Designer's message	Personality profile	Average ranking 1st- 2nd – 3rd	Significance level of Kendall's coefficient of concordance	Conclusion
Modern	Bright - Dim	B – A – C	Not significant	Sample did not reveal agreement
Lightweight	Simple – Complex	A – B – C	99%	Tower A is considered most simple (lightweight)
Elegant	Gentle – Violent	B and C – A	Not significant	Sample did not reveal agreement
	Moderate - Excessive	A – B – C	Not significant	Sample did not reveal agreement
Youthful spirit	Liberal – Authoritarian	B and C - A	99%	Towers B and C are the most liberal (youthful)
	Rebel – Conformist	C – B – A	99%	Tower C is the most rebellious (youthful)
	Optimistic – Pessimistic	B – C – A	Not significant	Sample did not reveal agreement
Joyful	Light-hearted - Serious-minded	C – B – A	99%	Tower C is the most light-hearted (joyful)
	Kind – Unkind	B and C - A	95%	Towers B and C are the most kind (joyful)
Flexible	Flexible – Inflexible	C – B – A	Not significant	Sample did not reveal agreement
Dynamic	Energetic – Unenergetic	C – B – A	99%	Tower C is the most energetic (dynamic)
Stable	Stable - Unstable	B – A – C	Not significant	Sample did not reveal agreement

Table 9. Analysis of the results of the survey on the personality profile of the towers for CD and DVD storage and verification of messages perceived from observation of the objects by the pannel of undergraduate industrial design students.

According to the findings obtained, the communication of a message of young spirit, dynamism and joyfulness were validated. Tower C (bionic tower 2) is the one which, according to the survey, more effectively conveys the desired messages, is considered the most dynamic, the most rebellious, most joyful and, together with tower B (bionic tower 1), most kind and most liberal. Regarding the transmission of the message of lightness, the personality profile related (simple - complex) did not translate so well the associated requirement. This might have led respondents to identify tower A (conventional) as the simplest, and therefore, according to the tenuous association, the lightest of the three. Interpretative meanings vary from person to person. The absence of actual experience of use of the towers on the part of respondents, who just exercised visual perception, may have also influenced and contributed to vagueness and lack of agreement among the respondents.

4.3 Validation of form optimization

The results concerning requirements contributing to the satisfaction of the goal of form optimization are shown in Table 10 (enhanced stability - according to force vector analysis), Table 11 (increased lightness - solid modelling analysis), Figure 2 (enhanced readability of CD titles - graphical depiction) and Figure 2 (scheme illustrating analytical stability modelling). For the first of the three requirements concerned by this goal, bionic tower 2 ranks in first place, while for the second requirement, bionic tower 1 is clearly the lightest, while for the last of the three requirements both bionic towers achieve a tie ahead of the conventional tower. The results support validation of the achievement of the goal sought of form optimization, albeit both bionic towers are deemed equivalent in this respect.

Maximum lateral disturbance to maintain stability	Conventional tower	Bionic tower 1	Bionic tower 2
At maximum capacity	49,96 N	49,59 N	74 N
At medium capacity	35,11 N	29,57 N	56,34 N

Table 10. Comparison of results for the maximum lateral disturbance tolerated without loss of stability in the three concepts.

Total mass	Conventional tower	Bionic tower 1	Bionic tower 2
Transport mode	11,049 Kg	4,763 Kg	8,087 Kg
Use mode	11,049 Kg	4,763 Kg	17,087 Kg

Table 11. Comparison of mass data among the three objects.

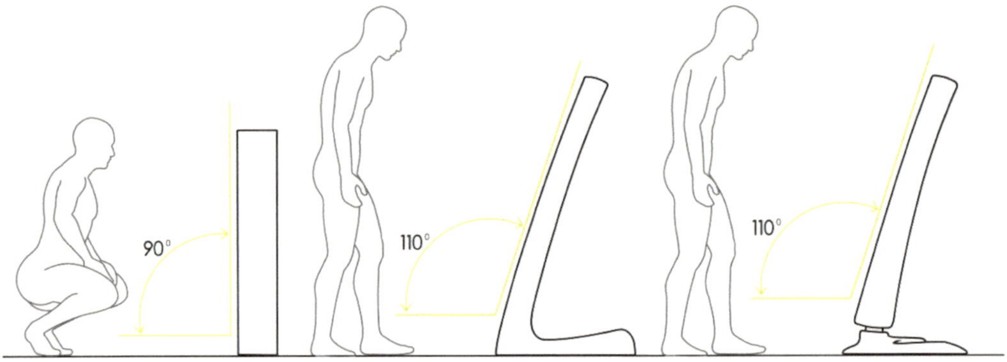

Fig. 2. Graphical depiction of readability of content titles for the three objects (from left to right: conventional tower, bionic tower 1 and bionic tower 2).

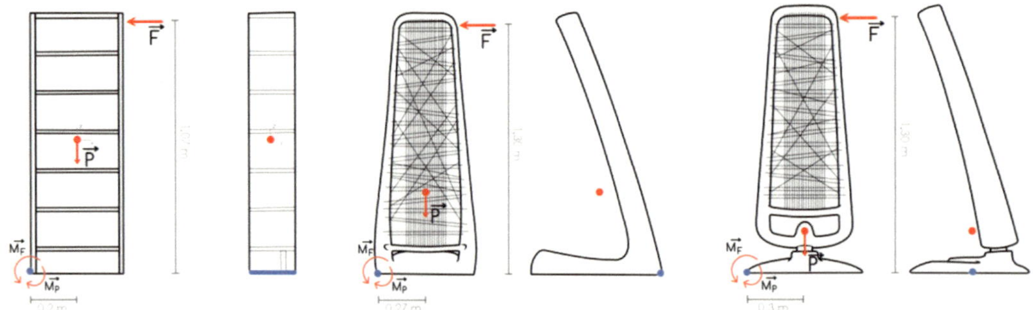

Fig. 3. Schematical depiction of analytical stability modelling for a lateral disturbance in the conventional CD rack, as well as in both the bionic towers designed.

4.4 Validation of organization effectiveness

To validate the goal of organization effectiveness, the proof of achievement of the requirement of storage with versatility of CDs, DVDs or books was sought by means of a graphical depiction (Figure 4) which is deemed self-explanatory with regard to this requirement's satisfaction.

Fig. 4. Depiction of three possibilities of dynamic storage of objects in the bionic towers.

4.5 Validation of paradigm innovation for improved functional performance

To validate the achievement of the goal of paradigm innovation for improved functional performance, verification of the achievement of the requirement of enhanced gripping of objects in the bionic towers was sought.

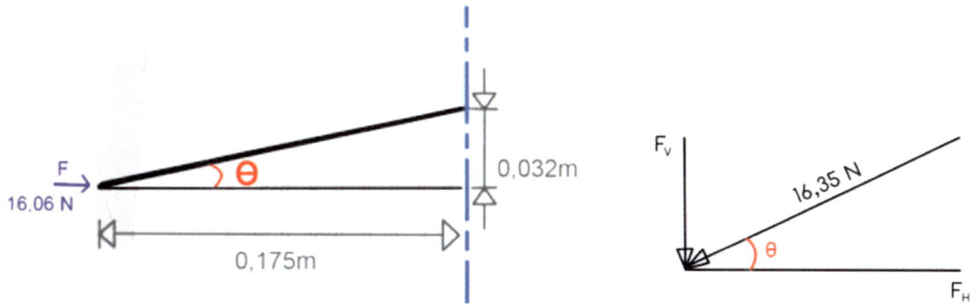

Fig. 5. Illustration of the changes in geometry of the web that generate object securing force.

Force and strength of material calculations were performed numerically and analytically, resulting in an estimation of approximately 0,5 N of vertical compression force per CD (based on analytical calculations developed from the physical comprehension of the phenomenon - Figure 5). Finite element modelling was pursued resulting in successful validation of the design for full capacity (Figure 6 – stress analysis under full capacity; Figure 7 – displacement under full load), yielding a safety factor of 166% and a maximum elastic (recoverable) displacement of 7.7 cm.

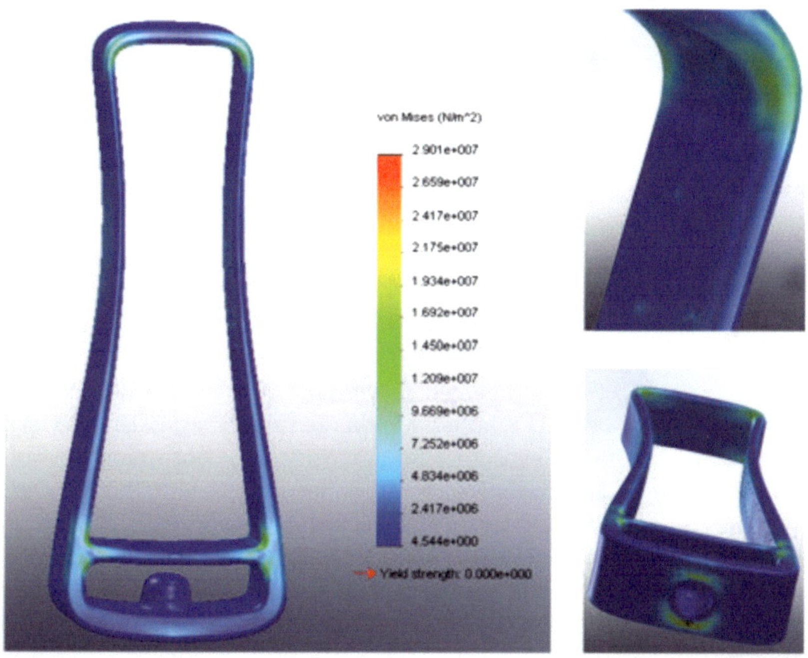

Fig. 6. Rendering of Von Mises stress analysis for bionic tower 2 frame under full load, obtained from educational 3D CAD software.

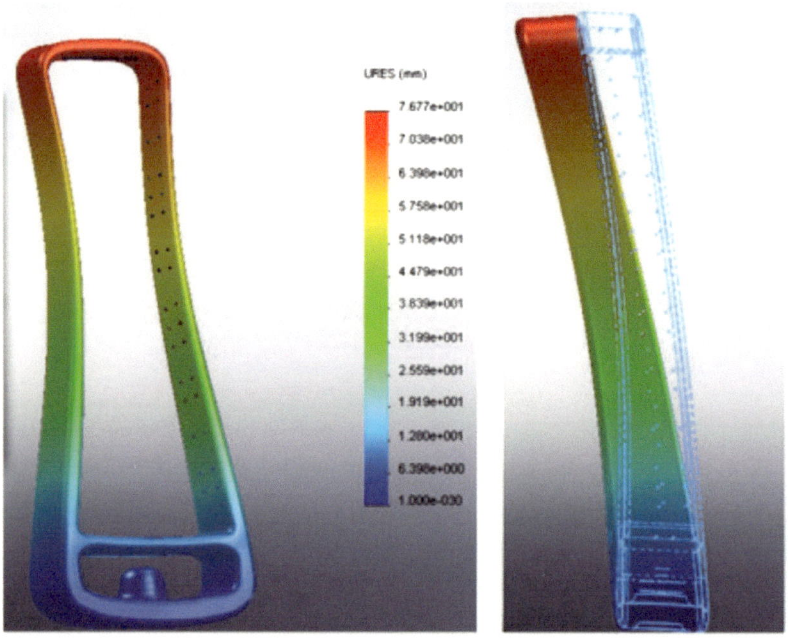

Fig. 7. Rendering of maximum elastic displacement for bionic tower 2 frame under full load, obtained from educational 3D CAD software.

5. Conclusion

From an industrial design perspective, the comparative analysis presented suggests that current methods to support bionic design, reported in literature, despite supporting specific goal satisfaction, are not effective across the whole spectrum of typical design goals. All the methods surveyed provide adequate support to the search for paradigm innovation, but form optimization, organization effectiveness and multiple requirement satisfaction are only adequately supported by some of the methods, albeit without any case of adequate support to all five goals found. Communication effectiveness is typically not supported in existing methods. The Aalborg method (Colombo, 2007) is deemed to adequately support attaining organization effectiveness, while the spiral design method (Biomimicry Institute, 2007) is deemed to adequately support attaining form optimization. Finally, the bio-inspired design method (Helms et al., 2009) is deemed applicable for problems seeking multiple requirements satisfaction, especially where trade-offs have to be established. Moreover, little support is given in the methods towards validation activities, concerning the satisfaction of the goals set for the design. The approaches to validation proposed in this chapter, combines engineering approaches with social science approaches to validation, in accordance with the nature of each of the goals focused. This validation process was demonstrated in a design case. Two variations of a novel bionic design for CD and DVD storage were designed using a combination of bio-inspired methods, and were then analysed in terms of satisfaction of requirements and validation of satisfaction of the goals sought, in comparison with a conventional solution for the same problem. This design case hence demonstrated the deployment of the validation process proposed in this chapter.

6. Acknowledgment

The research presented in this chapter was developed as part of the first auhtor's Master of Science thesis in industrial design engineering and as part of his ongoing doctoral studies, both supervised by the second author. A selection of results from the projects reported in this chapter have previously appeared in the conference papers Coelho & Versos (2010) and Versos & Coelho (2010), as well as in Coelho & Versos (2011) published by Inderscience and in Versos & Coelho (2011) published by Common Ground.

7. References

Altshuller, H. (1994). *The Art of Inventing (and Suddenly the Inventor Appeared)*, Translated by Lev Shulyak, Worcester, MA: Technical Innovation Center.

Benyus, J. (1997). *Biomimicry - Inovation Inspired by Nature*, New York: Harper Perennial

Biomimicry Institute. (2007). *Biomimicry Institute - Board*. URL: http://www.biomimicryinstitute.org/about-us/board.html (accessed December 29th, 2009).

Coelho, D.A. & Versos, C.A.M. (2010). An Approach to Validation of Technological Industrial Design Concepts with a Bionic Character, *Proceedings of the International Conference on Design and Product Development (ICDPD '10)*, Vouliagmeni, Athens, Greece, December 29-31, 2010.

Coelho, D.A. & Versos, C.A.M. (2011). A comparative analysis of five bionic design methods, *Int. Journal of Design Engineering*, special issue Design in Nature, pages pending.

Colombo, B. (2007). Biomimetic design for new technological developments, in Salmi, E., Stebbing, P., Burden G., Anusionwu, L. (Eds) *Cumulus Working Papers*, Helsinki, Finland: University of Art and Design Helsinki, pp. 29-36.

Hales, C. (1991). *Analysis of the Engineering Design Process in an Industrial Context*, Eastleigh, UK: Gants Hill Publications.

Helms, M., Vattam, S.S., Goel, A. (2009). Biologically inspired design: process and products, *Design Studies*, Vol. 30, No. -, pp. 606-622.

Hubka, V. & Eder, W.E. (1992). *Enfürung in die Konstruktionswissenschaft – Übersicht, Modell, Anleitungen*. Berlin: Springer Verlag.

Jordan, P. W. (2002). The Personalities of Products. In William S. Green and Patrick W. Jordan (Editors) *Pleasure with Products: Beyond Usability*, London: Taylor & Francis, pp. 19-48.

Junior, W., Guanabara, A., Silva, E. & Platcheck, E. (2002). *Proposta de uma Metodologia para o Desenvolvimento de Produtos Baseados no Estudo da Biónica*, Brasília: P&D - Pesquisa e Design.

Kindersley, D. (1995). *Máquinas Voadoras*, Lisboa: Editorial Verbo.

Lage, A. & Dias, S. (2003). *Desígnio - Teoria do design*, parte 2, Porto: Porto Editora.

Lloyd, E. (2008). *The History of Bionics*. URL: http://www.brighthub.com/science/medical/articles/9070.aspx (accessed January 10th, 2010)

Lovegrove, R. (2004). *Supernatural : the work of Ross Lovegrove*, New York, NY : Phaidon.

Pahl, G. & Beitz, W. (1996). *Engineering Design – A systematic Approach*, 2nd edition, London: Springer.

Pernodet, P. & Mehly, B. (2000). *Luigi Colani – Biography*, Paris: Dis Voir.

Roozenburg, N.F.M. & Eekels, J. (1995). *Product Design: Fundamentals and Methods*, Chichester: John Wiley & Sons.

Siegel, S. & Castellan, N. J. (1988). *Nonparametric Statistics for the Behavioural Sciences*, New York: McGraw-Hill.

Ulrich, K.T. & Eppinger, S.T. (2004). *Product Design and Development*, international edition, McGraw-Hill.

Versos, C.A.M. & Coelho, D.A. (2010). Iterative Design of a Novel Bionic CD Storage Shelf Demonstrating an Approach to Validation of Bionic Industrial Design Engineering Concepts, *Proceedings of the International Conference on Design and Product Development (ICDPD '10)*, Vouliagmeni, Athens, Greece, December 29-31, 2010.

Versos, C.A.M. & Coelho, D.A. (2011). An Approach to Validation of Industrial Design Concepts Inspired by Nature, *Design Principles and Practices: an International Journal*, volume 5, pages pending.

7

Product Design with Embodiment Design as a New Perspective

Lau Langeveld
Delft University of Technology
The Netherlands

1. Introduction

Product Design is the solving of a design problem from the assignment to the final product design. Many design methods may lead to a product design but also the design process which include embodiment design. The most time absorbing part of the design process is, in general, embodiment design: going from idea to realisation. In figure 1 a design process is drawn up, including embodiment design.

Product design is not business as usual, according to Kyffin (2007) of Philips Design. This statement has a presupposition that every product design needs an original approach of the design process to solve the design problem. Design needs to change, to enrich our cultures, to respond to the new world and new world economies, to create lifestyles, to develop quality life and explore innovative technologies in relevant new products.

First the answer should be given on the question "What is embodiment design?" Embodiment Design will be understood as the design phase where ideas get just some matter added and consequently a product structure, a product layout and a working principle. But the physical aspects should be meeting the requirement concerning the use, interaction and the ergonomics: the emotional aspects. During the design process the amount of uncertainty decreases and the amount of certainty increases. Every step in the design process is generating design information which reduces the uncertainty. The tools used in this design phase are countless.

Problems are enclosed in the assignment; at different levels and stages of design. Recognized problems can be transformed by the designer to a solution at certain levels and different stages of the product design process.

Every time a designer makes a choice, he also has to look at the method that will lead to the most ideal product design solution.

The goal of the product design is creating a product that fulfills its functions, looks beautiful, can be produced economically and is sustainable. The new perspective will be reached by the designer with methods that may bring embodiment design to the conceptual stage.

The key aim is to understand the alteration that is needed to move Embodiment Design to the conceptual phase. This is the new perspective for Embodiment Design which be researched in the near future.

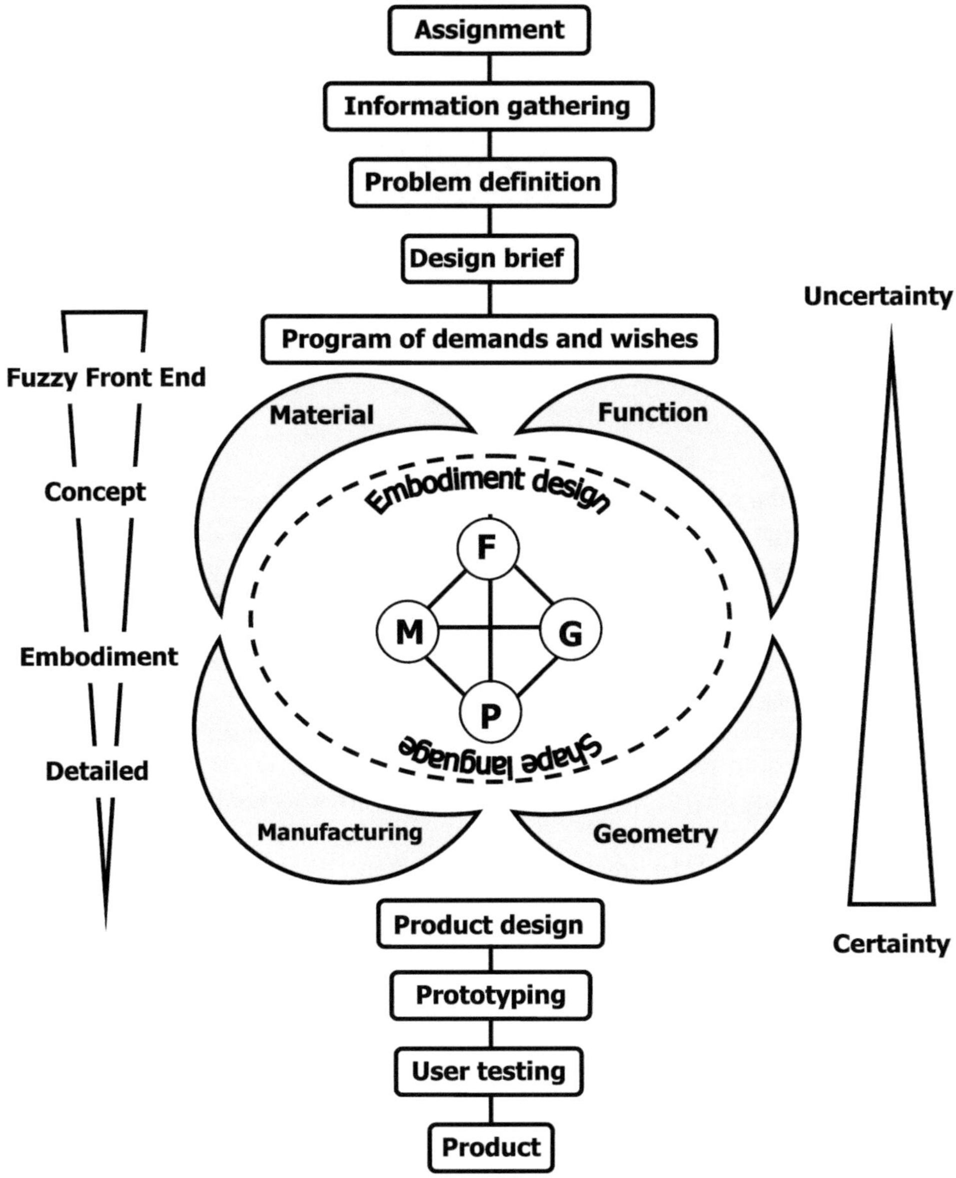

Fig. 1. The product design model, with the embodiment design model enclosed.

A product is assembled from components, sub-assemblies and parts into a working whole. The product has a structure with different levels. The product aspects can have constraints for certain parts that exclude parts form the product layout, for instance heating components in the area of plastic parts.

For a matter of fact all the product designs could be designed with the design aspects which may contain a number of embodiment design elements such as: design with X, engineering database, the designer needs, product structure, product layout, creativity, designer role, education, culture.

A designer has knowledge of technology and the design process. But the designer wants to develop himself by means of experience. Projects are excellent for the knowledge development and practice experience. Learning is inherent to the development of knowledge and experience.
Embodiment design should be educated in such a way that students realise this already in the early stages of the design process as far as the impact of the decisions is concerning on the embodiment design possibilities. During early stages could be taken decisions about solutions. These could be physical and emotional aspects such as: material, working principle, etc. which implicit determine embodiment design aspects. If the student is not aware of this then he limits the possibilities of embodiment design unintentional.
An engineering database has been developed to collect quantitative and qualitative information about products which stimulate the knowledge of products and experience with products at students during designing including the embodiment design phase. The stimuli come from incentive that is fed by the engineering database just during embodiment design. The three main goals of the engineering database are within the context of embodiment design education:

1. storing of design information
2. quick retrieval of information
3. the retrieved information should inspire the industrial designer

The embodiment design education takes place during the whole bachelor and master program of industrial design engineering. It is not an explicit course but integrated in many design courses in the bachelor and master program.

2. Problem definition

Product design concerns the solving of design problems from the assignment to the final product. It covers the whole design process. In the literature you find many design method, but every method has the same goals; a product design that works and solves the design problem. The design method to use depends on the design problem, the designer's knowledge and experience and the needs of the user. The main issues in the design process and embodiment design are creativity, innovation, performance fulfillment and high design efficiency. The design efficiency is an expression of assembly time and number of parts which is an index number how efficient the product design is done and how much design space is left for optimisation of the product design.
Embodiment design is a part of the whole design process, but in this phase of the process, the product concept gets concrete form and material, which includes certain manufacturing processes. But sometimes you want to focus on the main design aspects for instance implementation of a new material in a complete new business area. This means that embodiment has to take place already in the conceptual phase of the design process. The real problem is "how do you move embodiment design to the conceptual phase?" in an efficient way. This question should be answered in an ultimate manner, which gets gestalt in a new definition about embodiment design. Embodiment design can move to the conceptual phase by focusing on design with X, one of methods within embodiment design. The focus on the innovative product solution within embodiment design will make it easier to answer the question.
The idea of "the first time right" is also giving a new perspective to the whole design process and to embodiment design. Of course this includes the embodiment phase in the

new perspective by avoiding the iterations in the design process. This can only be reached when all stakeholders have the same goal in mind. This opportunity will seldom occur because collaboration requires compromises in a design team and in the organization, but it counts also for embodiment education.

3. Objectives

Product Design develops itself in a direction which may solve the design problem into product concepts and finally to a product design that will be prototyped and tested before starting a new manufacturing unit. The manufacturing unit may be equipped with a product oriented unit, a process oriented unit or a combined unit, a combination from product and process oriented unit. If during the design process the realization of the product was ignored, there is a good change that design iterations are necessary to enable an appropriate manufacturing method of the product. If embodiment design has moved from embodiment phase to the conceptual phase then the focus may be on designing or on making. This could be reached by looking for an innovative design solution. Time is often spilled by working on irrelevant and conservative design solutions also in the embodiment design stage. This spilling can be avoided by performing a good analysis phase in which mixed and innovative design solutions are found.

Design with X occurs to the new definition of embodiment design. But seeking for applications in a new domain can be done on focusing for innovative design solutions and in lesser view for the mixed design solution.

For a new perspective of embodiment design there is always one goal to be reached by going for innovative design solutions with more design in the solutions. It has to be efficient and creative manner to come up with design solutions.

4. Product design

Product design is the process of planning the product's specifications, according to Industry Canada, in their glossary of automotive terms.

Product design can be defined as the idea generation (Tassoul, 2009), concept development, testing and manufacturing or implementation of a physical object or service. Product Designers conceptualize and evaluate ideas, making them tangible through products in a more systematic approach.

Product design is concerned with the efficient and effective generation and development of ideas through a process that leads to new products, according to the book the fundamentals of product design (Morris,2009).

Product Design is defined (Walsh et al, 1988) as: the activity in which ideas and needs are given physical form, initially as solution concepts and then as a specific configuration or arrangement of elements, materials and components.

The above definitions and the one's who are still coming show the weakness in the realisation of the products and service but the designer should involve the stakeholders in the development of products or services to come to a good product including the embodiment design. Products are objects or services which are the results of designer activity. Before manufacturing, every part has to be detailed with, material, manufacturing and geometry. Additionally marketing, advertising, product introduction and distribution have to be done. Embodiment Design can help with problem solving and choices in the stadium of concept development (Otto & Wood, 2004). During idea generation it is to avoid irrelevant designs solutions and conservative design solutions nevertheless but focus on mixed design solutions and most on innovative design solutions, see figure 2.

Product design is the process of defining all the product's characteristics which affects product quality, product cost, and customer satisfaction. A product design could be designed badly than quality, cost and satisfaction go downward. Product design approaches regular need adaptation, because lifestyles change every decennium. Sometimes it is even going much faster for instance the use of computer, iphone, ipad, etc. So product design is not business as usual, because the process should be suited to the actual design problem. In most cases researching the design problem is the best way to start the design process. (Shih-WenHsio & Jyh-Rong Chou, 2004) The design team is often formed after the design research in which the boundaries are defined and all the design tasks that should be performed until the product is detailed and the prototype tested.

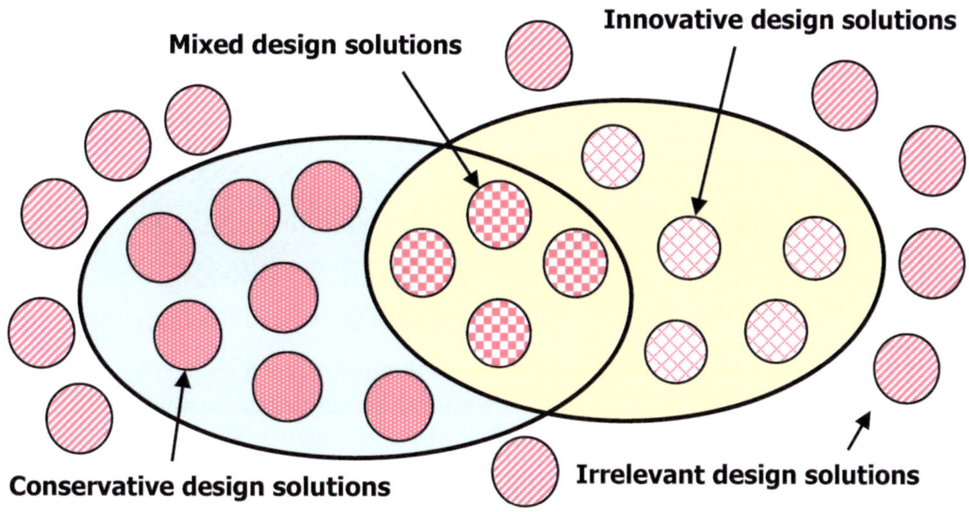

Fig. 2. Possible design solutions

The product design process contains at least the followings activities:

- Identification of 1market needs
- Problem analysis and formulation of the Design Brief
- Product Design Specification
- Concept development
- Embodiment design
- Detailed design
- Design for Assembly
- Life Cycle Assessment
- Evaluation

The above activities occur at different design levels and different stages. For instance evaluation should be done on the end of the every design stage. Competitors products should be researched for ideas which may be used in own product design. Here it is six of one and half dozen of the other such an important activity may be benchmarked, before starting the design process.

In figure 3 the product design model demonstrates the whole process from assignment to finished product. Embodiment design in practice is spread through a large part of the design process. The steps distinguished in the design process which in reality is not taken one by one, section 6 will explain more about this. Innovative product designs are realized by a creative process development which diverges from the classic step process.

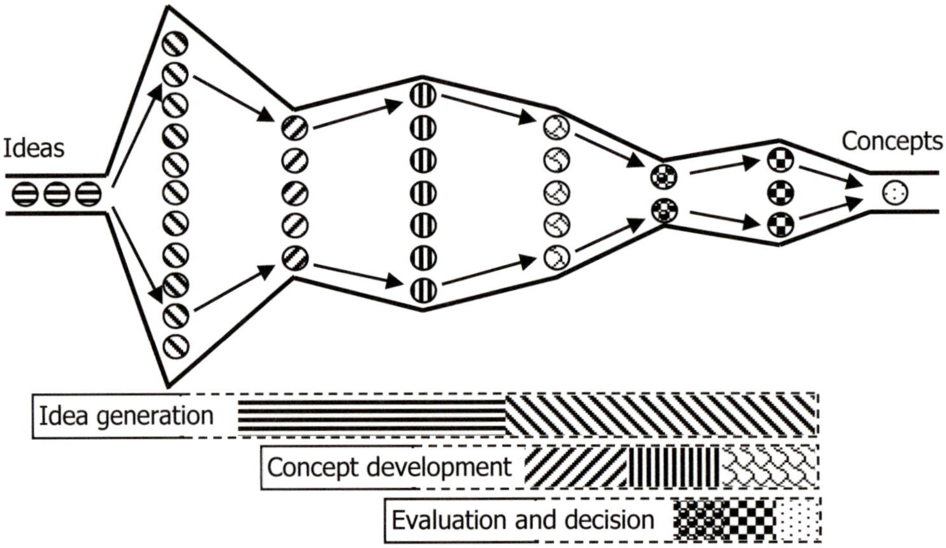

Fig. 3. The funnel model after Eekels and Rozenburg & Ullman and Eppinger

Concept development needs innovative idea generation, this includes the embodiment aspects. This bubbles up during the process from ideas to concepts. Evaluation and decision making are essential, because not all ideas can be developed into a concept. The funnel model is defined after Eekels & Rozenburg(1995), and also after Ullman & Eppinger.(2004)
The product presentation is a way to communicate with the prospective costumers. The embodiment of a product is an important part of this communication. There are many possibilities for presentation such as: projection, perspective, netting, time line, star, exploded view sketching, cross section, prototype, movie, 3D -views, e-drawing, etc. A nice example is given with 3D - view for housing parts which are just different on a small detail in figure 4.

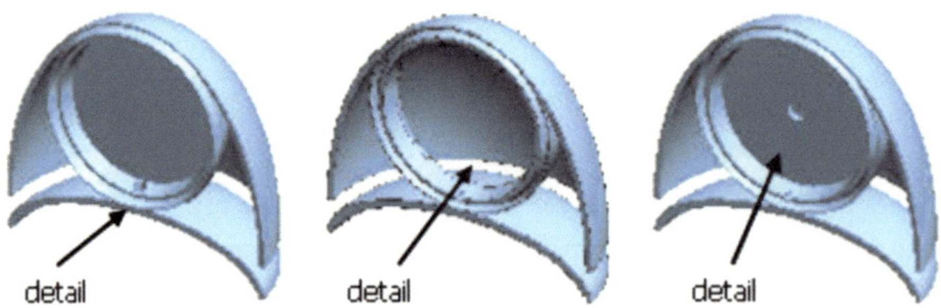

Fig. 4. Three housing parts in a 3D view with little differences in details

Product planning helps the company to realize the opportunities of product design and invests in the most likely product design. The product plan has to deal with the product development, product strategy, marketing, product portfolio which include embodiment design for each aspect. More specific, the product plan may be used to: product strategy and selection, defining target market or better competitive strength, distinguish from product competition, establish priorities in project development, high levels schedule, embodiment design, estimation of product cost and balancing product resources.
Embodiment is inherent to product design because in the design process you may take already decisions about embodiment during different stages of the design process such as conceptual design. Product planning is a process that runs parallel to the design process either embodiment design. Therefore preferably the planner and designer do the decision making together about the steps to make in the design process and in particular embodiment design. The planning process is necessary to identify and stimulate redesigns and new product designs with embodiment design as main concern, which can lead to new products. The goal of product planning and design should be avoiding investing in any chanceless product ideas. Market knowledge and product knowledge with a great share of creativity are necessary for the planning and design with a number of milestones just for the decision making on the right moment. The decision making must be on the right level in such a way that the decisions are not made about apples and pears.
There are still companies that have a formal product planning which lead to inflexibility in the embodiment design phase. Only the successful product ideas should be embodied, all the other ideas consume senseless time. This makes it hard to come to compromises in many cases.
The practical approach of embodiment design is shown in figure 5 with the physical main design issues of a product. This rocking chair is built with the material bamboo and is realized by transforming the bamboo in such forms that the parts can be used for the chair. The geometry concerns about the desired shapes, the interaction and the person who will sit in the chair, with are derived from a drawing or an explanation by the designer. In some cases, the designer is also the maker of the product.

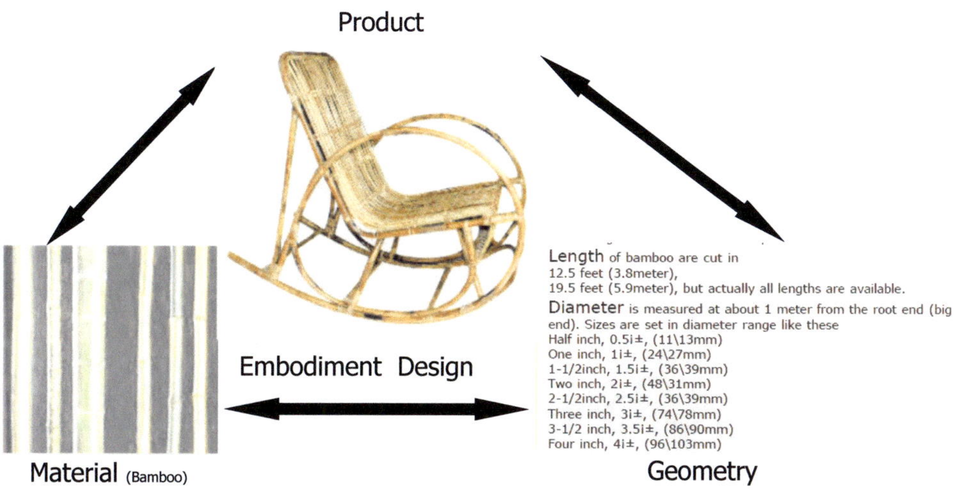

Fig. 5. Product, practical model

5. Embodiment design

Embodiment design is well known in product development. Kesselring (1954) was the first to refer to Embodiment Design and introduced a set of principles: minimum manufacturing costs, minimum requirements, minimum of weight, minimum losses and optimal handling. These principles are often calculated at the end of the design process and are typically used as verification.

The definition of embodiment design according to Pahl and Beitz (1996) runs as follows: "Embodiment Design is the part of the design process starting from the principle solution or concept of a consumer product. The design should be developed in accordance with engineering and economical criteria". This is a pure technical and economical consideration of Embodiment Design. But a product has more aspects than only the technical and economical ones. A product can also bring aspects about emotion, beauty, appeal and happiness the other values in live. People like to pay for these values if the earnings are higher than the cost of the basic needs.

The Embodiment Design phase is the part of the design process which is concerned about the production of the product concept, the engineering and the economical feasibility. The production contains the parts making and the product assembling.

However this doesn't open the new perspective of embodiment design. We propose a new definition of embodiment design and it runs as follows "Embodiment Design is designing with material, manufacturing and geometry to fulfill a new function or updating of the function". The emotional aspects such as: use, interaction ergonomics, etc. have to meet the requirement with the physical aspects.

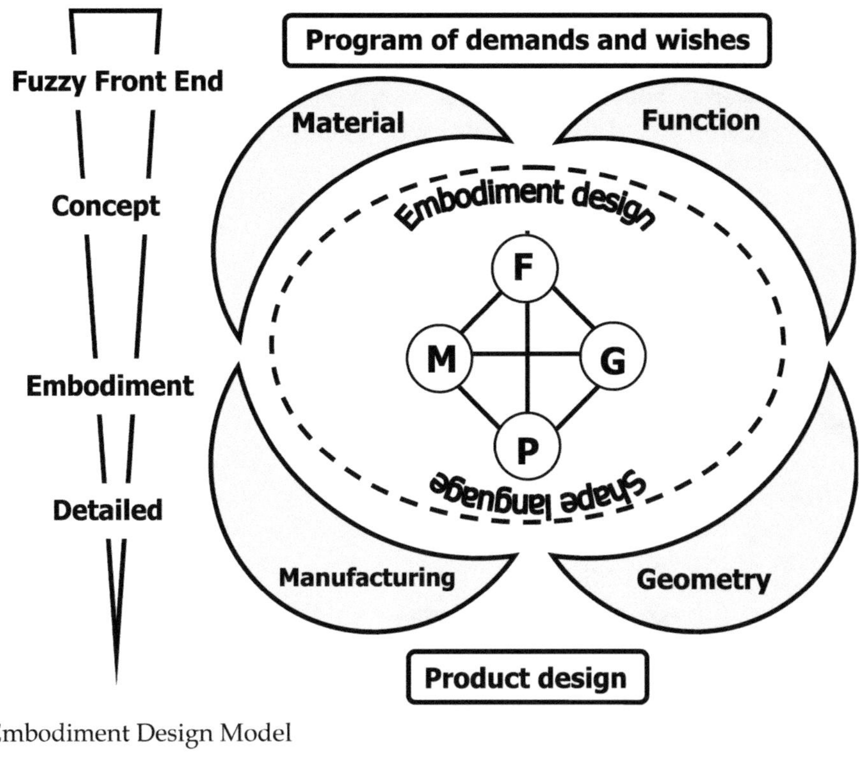

Fig. 6. Embodiment Design Model

Embodiment Design is giving matter to ideas, so a body is created in headlines, which will be detailed during the continuation of the design process.

The design aspects Function,(F), Material (M), Geometry (G) and Production, (P) in the FMGP-model have relations which are defined as design activities, see figure 6. These design activities can enrich existing products or product design concepts into innovative design solutions. The direction of an activity from the design aspects to a function is called Design with X.

All the product designs can be designed with these design aspects which may contain a number of embodiment design elements such as: design with x, engineering database, the designer needs, product structure, product-layout, the role of the designer, creativity, education and culture.

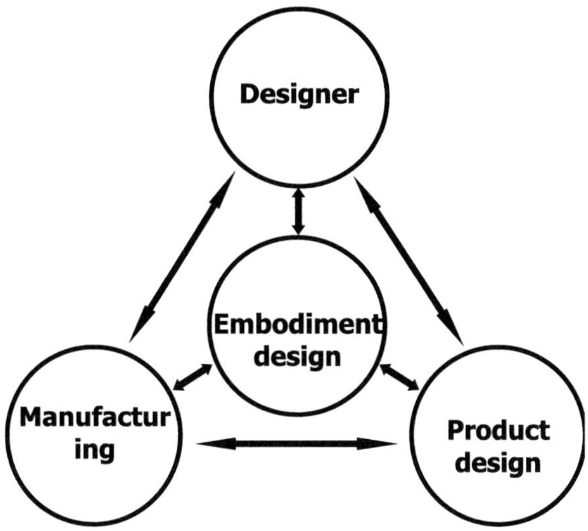

Fig. 7. The domains of making and doing

Embodiment design is a process of many different aspects in order to come to a product design. In figure 7, the domains of making and doing are provided which their mutual relations. Designers use embodiment design to follow a structured process, which depends on the design task.

The result of their doing is a product design which can actually be produced. The designers have to build their knowledge on manufacturing and even broader, on production. The product designs are related to the facility of manufacture systems and planning, the strategic and innovative aspects.

Embodiment design isn't an exclusive course, but part of advanced product design projects and other design courses in our bachelor and master program see figure 8. The assignments for advanced product projects are brought in by companies and institutes, so the 'design problems' are realistic. After the design brief has been formulated, the student groups start with embodiment design and finish the project with the testing of a functional prototype. Each group presents the results in the form of a report and a presentation for the other groups. The results must be in the area of a new working principle, cost reduction, new materials, parts reduction, use or other manufacturing processes, etc. These are all

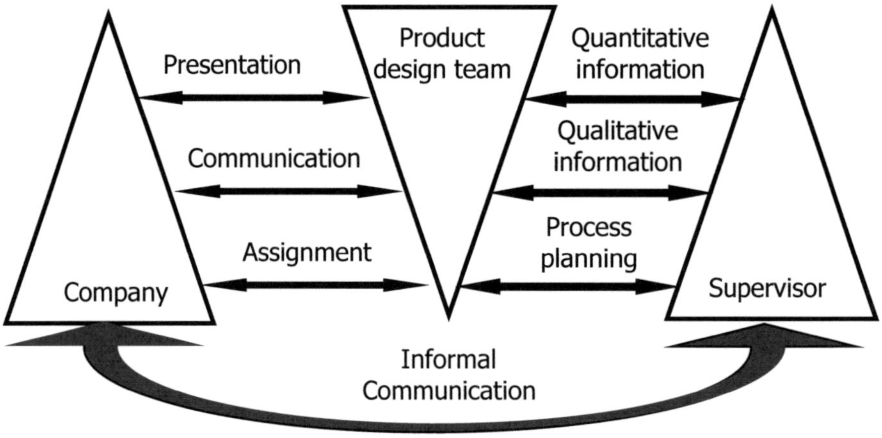

Fig. 8. The cultural difference between the stakeholders should be aware at project with companies

engineering aspect in which embodiment design can be of great assistance. Still, industrial design engineers have a tradition of simulating, calculating and testing the prototype.
A design aspect is an independent item that can be influenced by the other aspects.
Embodiment Design can start with the program of requirements and wishes of an existing product or concept design as shown schematically in figure 9. Either alone or in combination, the design aspects, material, function, geometry and process have to contribute to innovative design solutions. Designing with one design aspect for example process is called Design with Processing. The innovative design solution can be reached for 100% by process, geometry, material or function; that is called Design with X. Of course design with X could be design with anything however talking about design it is dealing with the main design aspects. This created an opportunity for embodiment design to penetrate into conceptual phase.
Ideas are transformed by a designer into bodies, which can be resulted in products, components and norm or standard parts. However the designer must avoid to do senseless

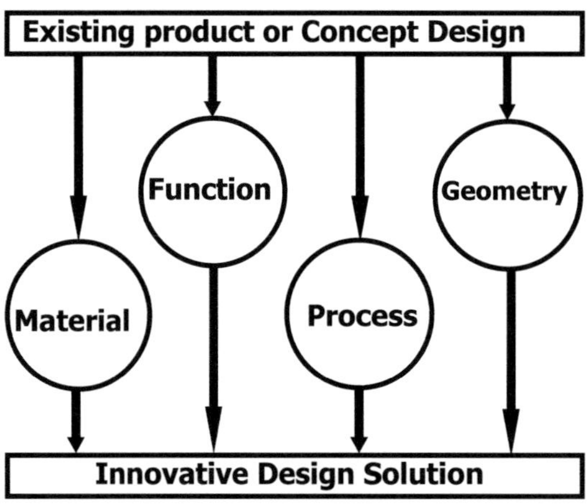

Fig. 9. From existing product to an innovative design solution

work so the designer has to recognize the irrelevant ideas and the conservative ideas and should not embodied them (figure 10) to bodies. The more the designer recognizes by doing how more efficient the design can be the designer is gradually getting more experienced. One idea may be realized into one body; this transition takes place in the head of a craftsman or artist, it could be art but it is craft. Embodiment design needs a plan that is not necessary for craft. Systematic design and engineering design occur but their results have lost the chance which lead to less solutions.

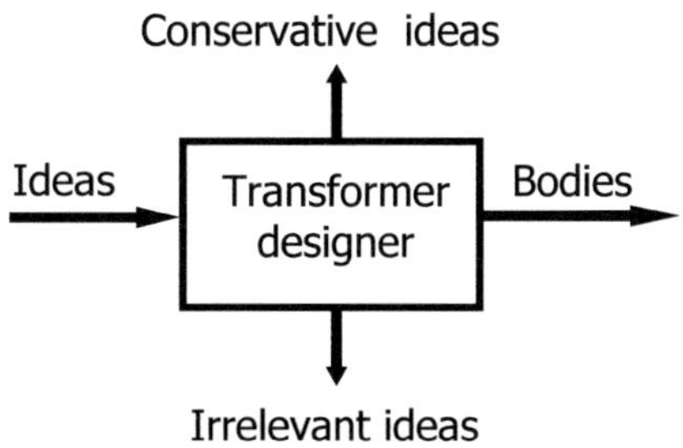

Fig. 10. Ideas transformed into bodies

No detailing takes place during the embodiment design phase. Detailing should take place at the moment that it goes over in engineering. In engineering, experience plays an important role. It is favorable to lay down the geometry, material and manufacture information on the technical drawing. For parts this information could be not sufficient, however the relational information may also necessary for instance for two moving parts.

Embodiment design could be brought to the conceptual phase, that could be led to more efficiency. Innovative design aspect could be taken as goal for the product design project with a main concern on embodiment design which leads to the innovative product design solutions. Stay to your design problem; do not run away to engineering and detailing. Then the design brings easier a product design solution which can be successful by using innovative design aspects or combinations. Innovations could also be done for items in the product organization or business financing. This does not have influence on the design of a product. Seek for the honest innovative design aspects, but concentrate on the design task because embodiment design gives a protected environment to do the design task on a creative way. The analytical approach of the design tasks influences the innovative power in the number of product ideas.

Decisions have to be made in the embodiment design phase at different moments in the process. In the Delft Design Guide (Boeijen & Daalhuizen, 2006) the following decision and selection methods are described: C-Box, Itemised Response/ PMI, vALUe, Harris profile, the Datum Method and Weighted Objectives Method. However for product design and part design no specific method is dedicated to product and design. We have filled the gap by creating a design decision matrix based on: who, why, what, where, how and when, see table 1. However the decisions and selections have indirect influence on the shape language. The design decision matrix (see tabel1) is a tool to identify how the design decisions are taken. It was inspired by the internet weblog learning journal from Lombardozzi, (2009) named Design Decisions.

Design Decision	Product Design	Part design
Decision making (Who)	Who decides which function could be fulfilled by product design? How might the product design be subdivided (but still a whole remains)?	By whom is the component or part being designed? What are the critical characteristics of part design to take into account?
Objectives (Why)	What are the goals and the performance objectives?	What are your goals and the performance objectives, concerning components and parts? How complex are the relation between objectives of the part?
Content (What)	What knowledge and skill areas need to be "covered"? What aspects of these topic's in the scope and out the scope, are important?	What information, procedures, skills, models, etc. will be shared with team members? What is in scope and out scope for the parts?
Delivery Method (Where)	Which will be the best delivery methods to accommodate the needed techniques, the overall preferences? What are the units' interfaces?	How are you going to deliver the part? What tools will be used to develop the part?
Techniques and Activities (How)	Which technique supports the best promotion of the product design?	What techniques will best contribute to the achievement of the objectives? What is a high level for part design?
Structure and Timing (When)	Which aspects of the product design need to be self-directive (pulled) vs. need to be instructed (pushed). How do you organize the research and design aspects of the product design? What is the intended time planning of the product design?	How do we organize the parts design in time? How are products broken down in parts? How do we represent the content (graphics, sound)? How long will it take to complete individual activities or components? What is the intended design time of individual parts?

Table 1. The design decisions matrix for product and part design

Within Embodiment Design the decisions are made by the designer, but at Detailed Design the part properties are established such as dimensions, material, tolerances, geometric tolerances, surface roughness and the volume (the number of parts in one product). In every stage of embodiment and detailed design the uncertainty will be decrease and hopefully come close to 100% certainty every time. Shape language should be used during this path from uncertainty to certainty, for the parts and product design. At part level most parameters are fixed and the manufacturer should deliver the part with the required performance so that the assembled parts will fulfill the estimated performance of the product. Detailed Design is the final touch of the product design and should be 100 % correct for the assembled product to fulfill the expected performance.

6. Product

Product are the red thread through embodiment design because the designer has to create something new, what it means for the designer need about product knowledge which he has to learn from existed products. Design experience shall do grow the product knowledge such as: material, manufacturing processes, aesthetic, interaction, use, appearance, etc.
A product structure can be made by tearing apart a product concept or an existing product to learn and gaining knowledge of products which be good for use in the embodiment design stage. The product will be split up in functional units, sub-assemblies or components, parts and raw material, see figure 11. In general, the raw material is not found in the product structure because raw material is another business than designing and manufacturing products. Design and manufacturing can take place also in different business units such as design studios and manufacturing plants, depending on business size, skills, quality, technical knowledge, total products costs. The business activity is only successful when all the ingredients are positive for product development.

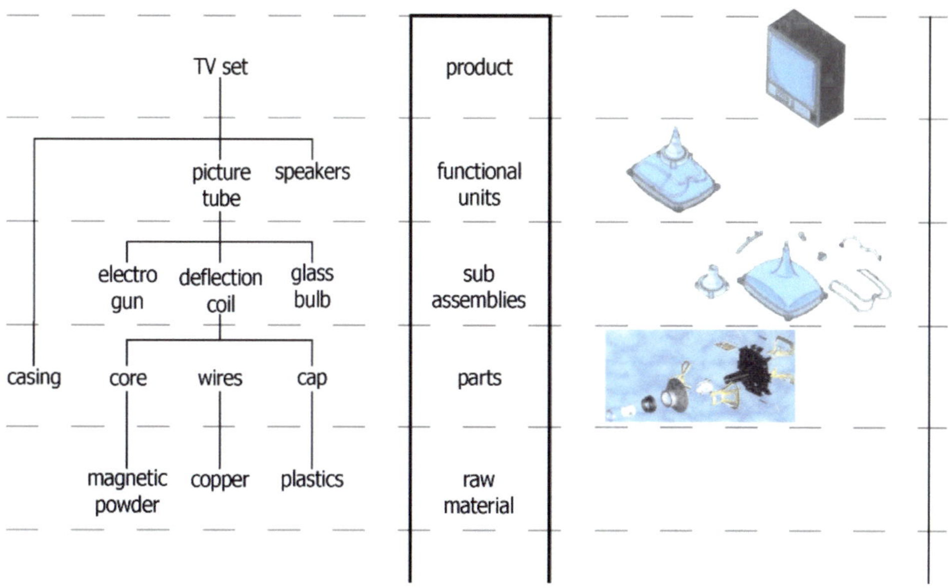

Fig. 11. Product structure of consumer product, which starts with raw materials

Product levels are acting as abstracting of the task in the product creation process. All product levels need their own approach; for example raw material require quite a complete other knowledge, skills and creativity than parts production.
Product layout arranges the assembly activity for manufacturing. The manufacturing may be organized in line systems of which the operations are sequential for a particular product.

The assembly system can also be organized as single point. Together they enable all kinds of arrangements, for instance a line system with sub-lines and a single point workstation.
Product aspects are identified as activities which are essential for a successful product. The hot topics for product aspects are sustainability, integration, embodiment design, healthy environment, feasibility, inter action, sound, etc. The topic embodiment design justify the possibility of making and the decisions are made in the uncertainty stage where it gets more certainty.
Time is money; however, the time available to create a successful product is always under pressure. But the "the first time right" principle is only possible if during the creativity and making processes, sufficient time is available for design development and manufacturing. Individual interests and short term successes are undermining the development of a successful product.
Product Management is about managing your products; manage what you offer to your customers at more or less fixed conditions. A product involves a more or less fixed outcome which encounters more or less stable, but at least pre-agreed terms to your customers. A product manager has the responsibility for the success of the product that means all the stakeholders will be served with the right tasks and budgets to get the product on the market in competitive way. Embodiment design is the base for an efficient product design process, but the product management should be on the same quality.

7. Design

Ulrich and Eppinger (2004) assign the next meaning to design: the design function plays the lead role in defining the physical form of the product to meet costumer best needs, these are also the embodiment task. The design function includes engineering design (mechanical, electrical, software, etc) and industrial design (aesthetics, ergonomics, user interfaces, etc.).
Design changes the world, however, before that happens many strategic product development decisions have to be made and the bodies are getting the form and geometry of the parts and product with embodiment design. But the design on itself does not change the world. The designer has the human instinct that every problem could be transformed to a design problem which may enclose a design revolution solution.(Pilliton 2009) mentioned in the book "Design Revolution". A great number of design solutions that have changed the world in different areas such as: water, well being, energy, education, mobility, playing, enterprises, etc.
In the innovative Design Solution model of an existing product or concept design, the main design aspects can adopt the innovative character of an aspect or combination for embodiment design. The goal will be a new product design that is needed on the market. For innovation in application of materials, the easiest way is to search for successful design applications, because experience with innovative material use starts at that moment.
The trade-off between design time and money evokes questions such as "how much time is available for Design?" The available design time should be used efficiently, also for embodiment design stage. The time management is a prerequisite to put the products to the market in time and within the budget. Time management involves time to market and time for designing and embodiment design for a good design. You have to plan and manage the whole embodiment design. It is necessary for the overall process as well as the design process. Time management costs time and money. However, a good balance between financing and designing is a must for a company. More design time may lead to better

designs because the solution can be considered in more depth. However, the deepening should be done on design aspects that request this, for instance deepening of manufacturing process by mechanical process modeling.

8. Designer

The designer is a human being with the ability to create products which the market needs with embodiment design in mind. The industrial designer needs input from art, social, cultural and technology knowledge and experience to be able to make a product design see figure 12.

Furthermore, the designer should have creativity, inspiration, motivation and emotion. These qualities enable an energized transition which can be done by the industrial designer who is using embodiment design aspects that may lead to a predictable product design.

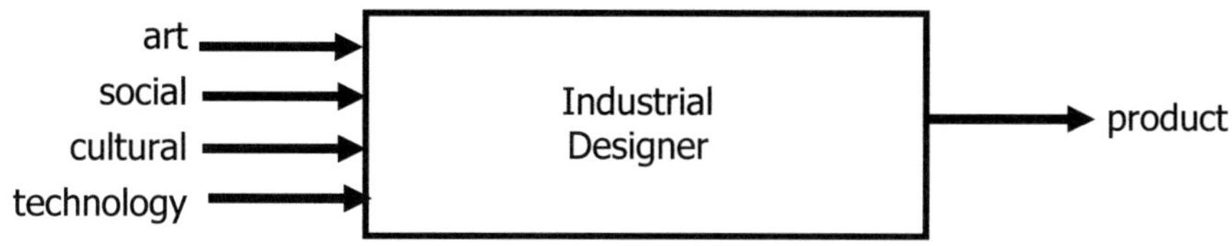

Fig. 12. Black box model of the industrial designer

The fundamental needs of the designer are conditioning and contact. For every need, we can distinguish three levels of interaction, which are shown in table 2. These needs are expected from the prospective designer during his or her education. Embodiment design demands all relational needs to meet on a well designed product with high expectation, these relational needs should be fulfilled.

	conditioning	contact
biological	Need of expression by sketching Need of making	Search for design project 3-D-Model making
psychosocial	Need of being designer	Need to contact other designer
existence	Need to design consumer products	Search to sense of design

Table 2. The matrix of the designer relational need (after interpretation of Nuttin, 1984)

These are all design elements which bring the designer in an environment where he can create the innovative product designs in the embodiment design stage. Knowledge, experience and skills are the qualities of the designer which are fed by personal designer needs, inspiration which stimulates creativity; structure of design process plays, an important role in the designer life, education of developments on different design areas and the culture of living and designing. The hierarchy of designer needs have also been defined on Maslow (1970). It starts with the basic needs (knowledge, methods and tools) and the highest need is design research, see figure 13. Nowadays embodiment design is used on conceptual design level so the designer has to follow education that trained on the level of acting. The prospective designers have to develop themselves to fulfill the needs of concept design, which is mostly found in design programs at university level.

Fig. 13. Hierarchy of designer needs based on Maslow

The industrial designer can follow a certain path of education at a design school. During the educational program the industrial designer typically starts with the basic knowledge and progresses step by step in the hierarchy of the designer needs. An autodidact can follow these steps in his or her tempo, but gaining an overview of the right competences from each step costs much more time. Nowadays, the advice is to take a design course or apply for a design program, which is a more effective for the development to be a designer.

Embodiment design is mostly not offered exclusively but in the form of design courses with accents on embodiment design for example Advanced Design Project in master program of the industrial design engineering school in Delft.

The prospective designer applies for a degree in Industrial Design Engineering. He follows the program and has certain expectations of design education. These expectations are based on three main design capabilities: knowledge of design, communication of design and training, see figure 14.

Sometimes there is a great difference between the design program and the expectation of the prospective designer. If a selection is necessary by abounded of application, the offered program will match better the expectations? The applicants are more motivated for their enrollment and have the opportunity to their skills. However, the designers have to form themselves during the design projects which they are following in the bachelor and master program.

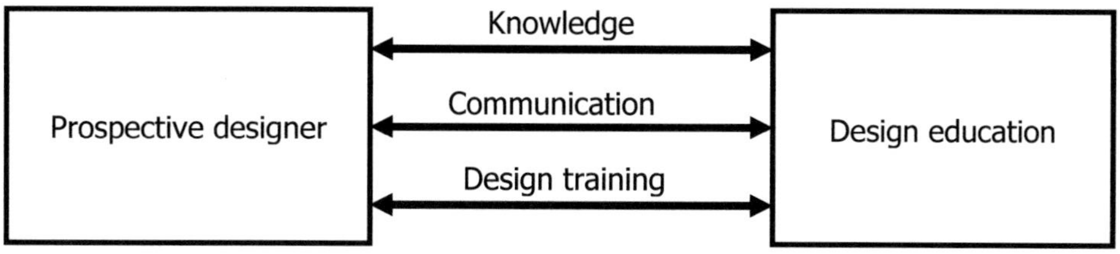

Fig. 14. The relation of the prospective designer and design education at his institute

9. Engineering database

In the embodiment design stage it is a possibility to orientate on the design problem to solve the design problem with knowledge from an engineering database that can inspire the designer.
The engineering database has three main goals within the educational context:

1. storing of design information
2. quick retrieval of information
3. the retrieved information should inspire the industrial designer

In design education enormous quantities of design information are generated in all kinds of courses and design projects. This information can be useful as an inspiration source for future design assignments. Thus, storing this design information is practical, if it can be quickly retrieved. Quick searching of design information is a need, if a designer wants to use the information for fulfilling a new design task. The data processing should be fast and reliable. For this reason, the input of product design information must be correct. This intensifies the need for structure and unambiguous data in product design information.
The stored design information must be reviewed regularly to check whether the information is still up-to-date. The size of the database may be large; however, it is only useful if the information is up-to-date. The search time should not be too long, otherwise the chances of inspiring the designer will decrease, which is shown in figure 15.

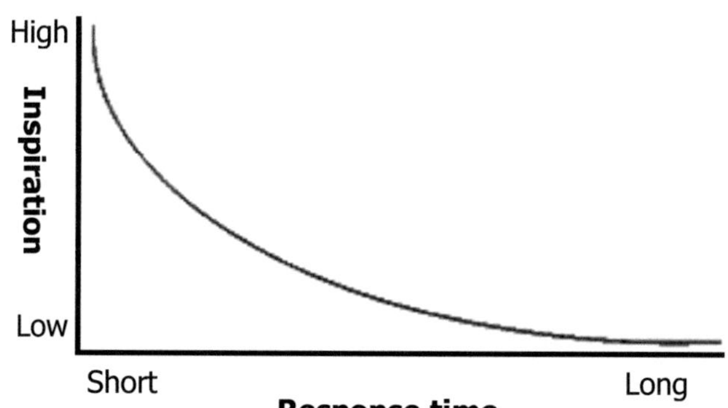

Fig. 15. Curve of the response time versus inspiration

The design education operates under high pressure, because the right information should be chosen from all the information offered for a specific design task. The present design assignments are so extended that no time could be spent on searching for the necessary information. Sources like the internet can deliver some information, but the amount of data is often overwhelming. Each day it is getting harder to find useful information on the internet. The engineering database can also fill up quickly.

Finally, the use of an engineering database in design education can prevent design students from aimless searching on the internet or in other information sources. The goal is to allow designers to find the needed information in the database, so that he or she is inspired. The engineering database can be consulted for inspiration in all phases of the design process. Inspiration gives motivation and inspiration challenges the creativity and inspiration challenges the creativity. The inspiration should be conceptual inspiration and detailed inspiration, see figure 16.

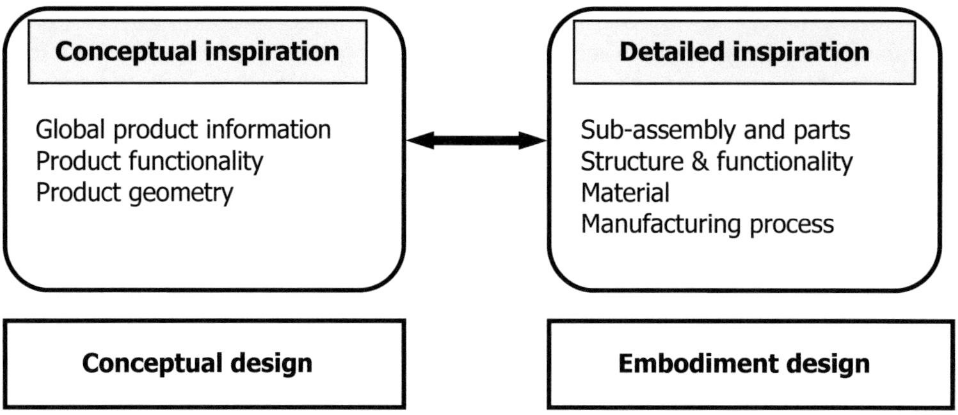

Fig. 16. Conceptual and detailed inspiration

Products have a structure that describes mostly how they are composed, with the relations out of the FMPG- model.

The product structure, figure 17, distinguishes several levels: product level, sub-assembly level and part level. The FMPG model is applied at the part level, but the design aspects change at component and product level

Components and sub assemblies can be bought or outsourced, depending on company strategy, volume on year base, uniqueness. Components are not made in the fabrication place itself. Sub-assemblies can be made inside manufacturing plants or produced on record.

Parts are manufactured in special shops to keep the cost to a minimum. The biggest part makers in the world are India and China, but they are also under pressure about the labour costs. Shipping the manufactured parts towards the place where the assemblies are done is relatively cheap in spite of the large distance between the part manufacturer and the assembly site. Parts which are assembled can be manufactured separately. So manufacturers of parts and sub—assemblies can be found all over the world.

Standard and Norm parts are produced in mass production; in this type of production, nearly all labor cost is cut away. The production of washers, for example, can be done almost fully automated.

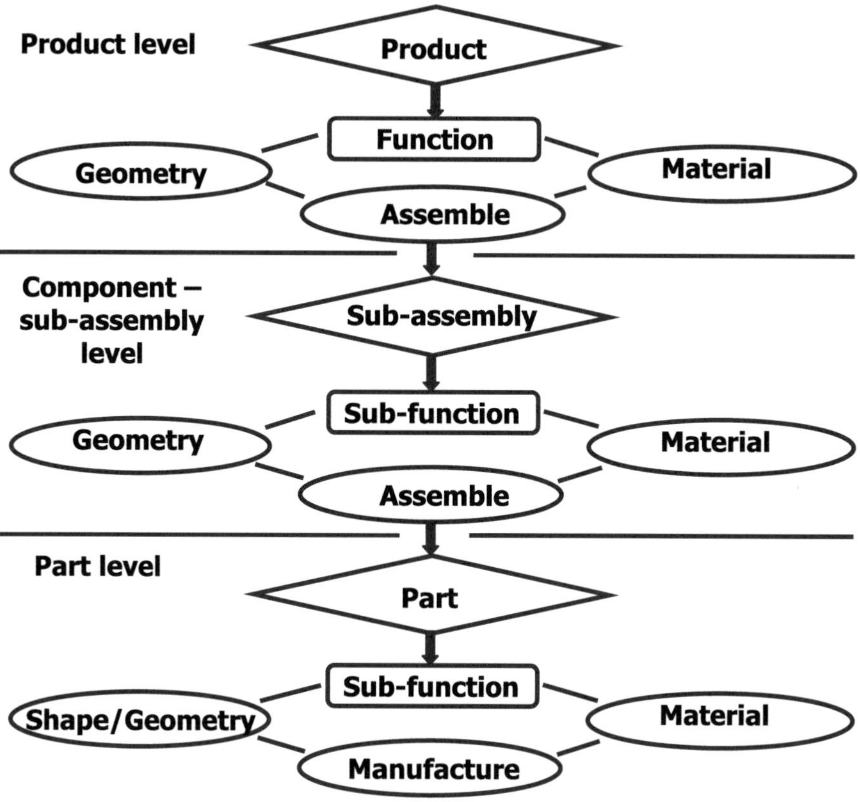

Fig. 17. Product structure on product, sub-assembly and part level with the design aspects

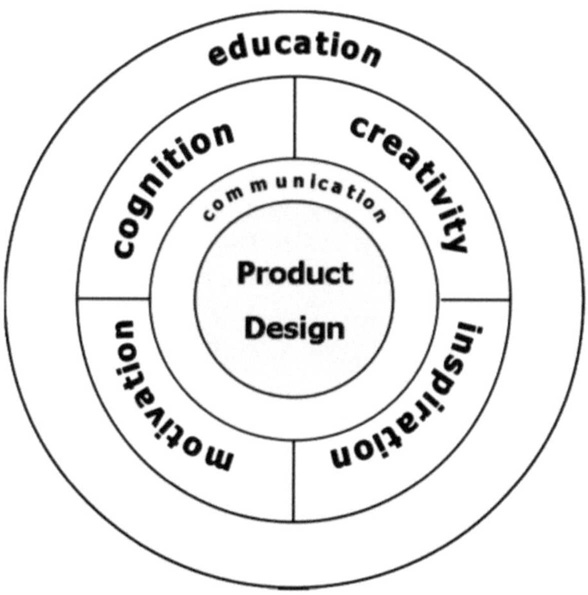

Fig. 18. Product Design Education model

Obviously modeling in 3D-systems comes up with views, sections, rendering as expression in visual images but e-drawings are also possible. This means walking around the product and check out or it is visual correct or wrong.
Calculation on parts, components and products may inspire the designer to optimizations. This is useful detailed inspiration information that is retrievable in databases.
Estimation of costs, market potential, appearance, color, etc., are necessary to become from uncertainty to certainty at the moment that every aspect is detailed and the production may start at this moment. But embodiment design is always searching to a solution that comes closer to certainty in spite of the uncertainty. Estimations help to increase uncertainty until it fits within the bandwidth that gives confidence about the solution area.

10. Embodiment design education

The education product design model shows the condition how an institute may arrange an education program to Industrial Design Engineering on Bachelor and Master Level, see figure 18. The cognition, motivation, inspiration and creativity are main properties of the product design program which should translate into design courses and domain courses which should support the design courses. Embodiment design is covered in divers design courses of many design programs in some design courses lay the accent on different main aspects of embodiment design.
Shape language is such main aspects just as a part of the communication which has to be taught in all product design courses. Cognition is more used by courses which are engineering orientated such as product in motion, technical product optimization, industrial production, technical modeling, etc.

Fig. 19. Cycle :model of creativity and inspiration

Creativity and inspiration are stimulating each other, as shown in the cycle model in figure 19. It is possible that no creativity may lead to less inspiration, what means to stop for a while with designing until the inspiration is coming back. Motivation comes from different areas principally with a lot of energy. This energy is picked up by designing with inspiration and creativity. Motivation or better feeling well is a must for a designer in the embodiment design stage. Education should focus on conditioning of the designer and offer all kind of tools for the embodiment design phase so the motivation stays on right level. The communication should lead to motivation and creativity of the students, because the knowledge they have to learn and to create by their selves.
A product design is the result of communication between a designer and his client, only good communication will lead to error-free product designs. Embodiment design is a part of product design with a new perspective that needs some revision for design projects. The design projects and courses, with the focus on embodiment design, should be further

improved, because embodiment design should be learned efficiently and also experienced by applying it in the new perspective.
Embodiment design will also be communicate where the technical engineering language plays a connected role in the whole. The shape of a product should be communicated by different designers which may lead to misunderstanding but if they have the same goal then it leads mostly to success.
The communication goes verbal and visual; every designer has his own means to communicate his design work.

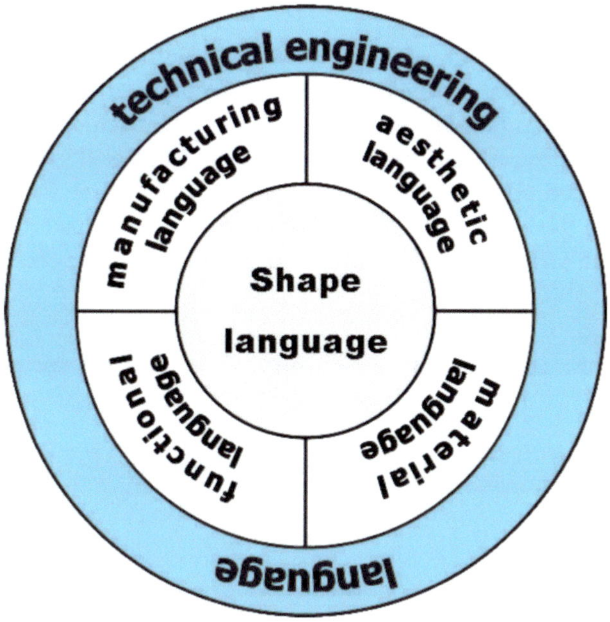

Fig. 20. Technical Engineering Language model which contains shape language

Communication is a way in which product designs respectively embodiment design become their gestalt by means of technical engineering language see figure 20. We need a language to transforms ideas into product designs. Within this technical engineering language, we can distinguish four more specific domain languages:

- Functional language
- Manufacturing language
- Aesthetic language
- Material language

Manufacturing language contains specific manufacturing terms. For instance volume is the number of pieces made in one batch; yearly volume goes about the production in one year.

Functional language indicates with specific terms how a product can fulfill its physical principle. The specific term contains always an activity which fulfills the product function or sub function, for example, coffee making.

Aesthetic language expresses with specific terms how one experiences the form and what the form does with you. For instance, a natural form may please you quite well, because of recognizing the beauty in nature.

Material language is giving specific terms about the physical properties of the material and how you can make and assemble the parts into a product.

Shape language has the specific terms to express the shape in a general way of material, function, manufacturing and aesthetics in the engineering way.

Realization of functions is always engineered by physical appearance; however, emotion can play a role, too.

During the embodiment design stage should be practiced the technical engineering language to come to a successful design solution. Every kind of education is lagging behind the facts in industry, including academic education. When education does not introduce innovation in their design projects and courses, the students will get far behind the reality when they get graduated. About embodiment design is it the same. Poor education does not have the capability to develop any education program with a focus on embodiment design, design projects that stimulate the student to realize an innovative product design.

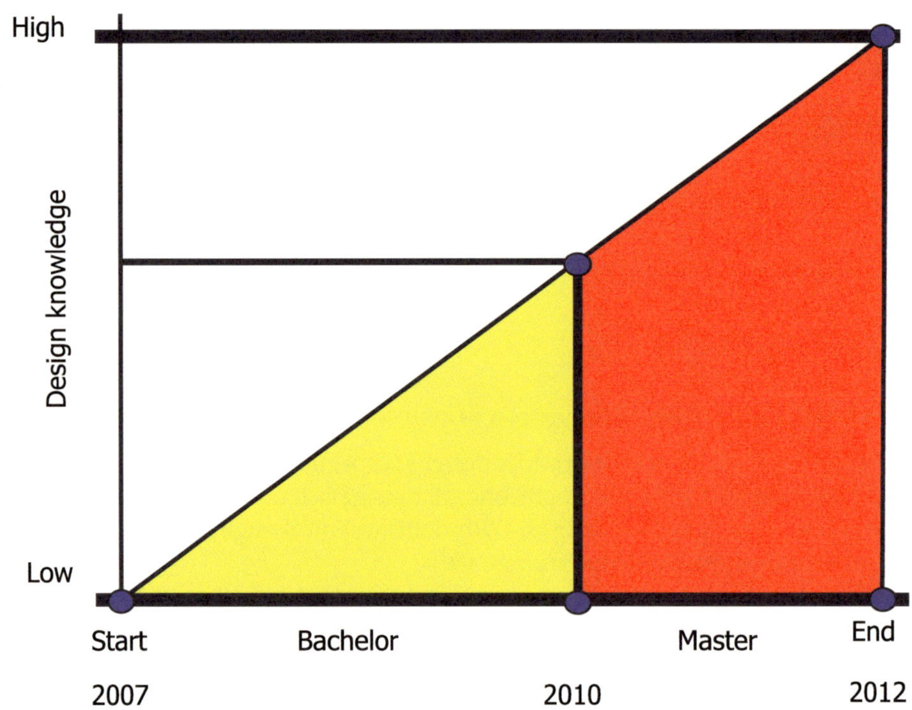

Fig. 21. Design knowledge during educations programs

A design staff must recognize the innovative product concepts which are elaborated into a product that has an innovative character, also in the embodiment phase. Here you need the use of embodiment design in the new perspective. The academic Industrial Design programs, at the moment, have a bachelor and master structure. At the end of the bachelor the design knowledge including embodiment design should be on a sufficient level for a bachelor to look for a job or do the master, see figure 21. The design bachelors should have the capacity to develop their own identity from junior designer to senior designer. The design knowledge has to be the foundation for the independent identity development. It is only one pillar of the successful Industrial Designer, because creativity and social skills are also essential elements to build on the designer to be as a personality. All this could be trained with embodiment design. The design schools have to offer design programs that give the possibility to be conspicuous in industrial design. This does not mean that the other regular design programs are not good enough. Attention should be paid to embodiment design in the new perspective just for opening perspectives for creating of the design identity.

The master program should not be more of the same, but provoke design knowledge that motivates and stimulates to do excellent design projects and courses. This design knowledge is gained from the results of design projects with predominantly industry design problems. This offers new perspectives for embodiment design, but also for product design. The real knowledge generator should be identified and developed to an education development system. Valuable knowledge should not be thrown away, even if it is not directly science knowledge.

11. Discussion

Product design includes embodiment design, but this is the strength and at the same moment the weakness of it. Embodiment design is included in the design process which is from assignment to the final product design. In the embodiment design phase the ideas get the physical form by the transition of ideas into concepts by the industrial designers. However the emotional contribution should be also transformed into the physical form. The physical form is an extension from material, geometry and manufacturing. Design with X is useful for many design processes with the goal of a product design. To promote a new material, first an amount of mechanical, electrical, thermal properties have to known or tested before you used the method successfully. The design process can take place on so many manners: sequential, cycles, spherical, random, etc. So the designer chooses the design process that fits the best for solving the design problem. All the processes have a moment that ideas have to transform into preliminary design, the first stage of embodiment design and after that detailing takes place. The new perspective brings embodiment design to the conceptual phase of the design process. When the designer wants to do the design effectively then a concentration on one of the main design aspects may be applied with design with X. It may happen that design with coffee making, could be solved with an innovative solution for example the Senseo coffee maker from Philips, a design with X solution.

With design with X conservative and irrelevant designs solutions can be avoided, they do not show up in number of possible solutions. If the designer looks for innovative design solutions then there is no space for the other design solutions. Focusing on one main design aspect helps to solve the design problem in an efficient way.

Decision making is an embodiment design activity that may be pushed forward during the embodiment design process, but if the decision is taken on the right moment in the process then the result gives a new starting point to gather information for next goal in the design process. The design decisions have to be made on who, why, what, where, how and when for product and part design. The questions are very helpful to support the decision making for that specific area. It provides a new perspective to embodiment design; it is a tool that has to be experienced by using the design decision matrix during a number of design tasks.

A product has a structure which is depending on the number of parts, components or sub-assemblies and units. The levels take care of items for that specific level. However design research on embodiment design, product management, design time, layout and feasibility are efforts that help to indicate the new perspectives for embodiment design and product design. These aspects are mostly in the area of organization and technology, where the creating process (Prabir Sakar, 2007) is supported with creativity as main driver.

Design is the most difficult phenomenon because it means everything, but time and products are concepts that give content to product design. The content may be innovatively, quality for money, beauty, sustainability, etc. The context of the product design problem has the power in itself to make a transition from ideas to a product design by an industrial designer.

The designers should have a certain aim of life for their eyes; creating products that make the life comfortable. The designers have a need that is depending on their education and capability. It may be on different designer need levels according to Maslow. Education can only support the need of the designer in his development, the industrial designer he wants to be. The support contains design training, communication and design knowledge. All the called aspects should be optimized for the junior designer. Experience in design does grow with experiences so they may grow out to a senior designer. In daily life the designer generate new knowledge, design ideas, product designs, part designs, etc., all valuable information If this is collected in a structured engineering database then it is shareable with all designer world wide. The base for such database is mentioned as engineering database with an ordering that can inspire the designer by means of the designs examples on conceptual and detailed level.

All the generated design information also the abounded information after a decision can be useful for sharing it. So the valuable information of alternatives principles, ideas, concepts and product designs have to be judged before putting them into in the database.

12. Conclusions

Embodiment Design is a part of the design process with a new perspective that is expressed in the new definition: "Embodiment Design is designing with materials, manufacturing and geometry to fulfill a new function or updating of the function", but the emotional aspects of the product should be met the requirements about: use, interaction, ergonomics, etc. Embodiment Design is giving matter to ideas, so a body is created in headlines, which get

detailed by continuing of the design process. This takes place on different levels which depends on the place in the design process.
Particular embodiment design aspects of an existing product or concept product design should be avoided iterations to come to an innovative design solution. Right choices occur not always for hundred percent, this will still lead to iteration as compromise of the design aspects.
Designing and decision making are the main activities of embodiment design but decision making could only be done by means of feasibility studies, estimation of cost, implementation of ergonomics data, and physical calculation such as; strength and stiffness, sound intensity, thermal effects, etc. However designing is using the designer skills to solve the design problems which are formulated in the design brief.
Design decisions have always to be taken on the right time, on the right place, and with the right knowledge. However the design decision should be taken in the right stage of the embodiment design phase. But time pressure disturbs a good design decision, because the designer can not research sufficiently all the aspects of design decision. If you get enough time to analyze the design problem then embodiment design can prevent iterations based on the principle "First time right"

13. References

Ashby, M.F., 1999, *Materials selection in Mechanical Design*, Butterworth Heinemann, ISBN 0 7506 4357 9, London, 1999 second ed.

Boeijen, A.,Daalhuizen, J. , 2010, *Delft Design Guide*, TUDelft courseware, ISBN 978-90-5155-065-5, Delft

Kesselring F., 1954, *Technische Kompositionslehre*, Springer, Berlin

Kyffin, S., 2007, Keynote: *Shaping the Future*, 9th International Conference Engineering and Product Design Education, Newcastle on Tyne

Lombardozzi,C., 2009, *Design Decisions*, Learning Journal, learningjournal.worldpress.com, posted 2 November 2006.

Morris, R., 2009, *The fundamentals of Product Design*, AVA Publishing SA, ISBN 978-2940373-17-8, Singapore

Nuttin,J., 1984, *A relational theory of behavior dynamics*, University of Leuven, Leuven

Otto, K., Wood, K., 2004, *Product Design*, Pearson Education, First Indian Reprint, ISBN 81-297-0271-1, Singapore

Pillito, E., 2009, *Design Revolution*, Thames & Hudson, ISBN 0500288402, New York

Pahl, G., .Beitz, W., 1996, *Engineering Design, a systematic approach*, London, 1996 second ed

Prabir Sakar, 2007, *Development of a support for effective concept exploration to enhance creativity of engineering designers*, Phd Thesis, Centre for Product Design and Manufacturing of Indian Institute of Science, Bangalore

Roozenburg, N.F.M., Eekels, J., 1995, *Product Design, Fundamentals and Methods*, Wiley, ISBN 047 1943517 1943518, Michigan

Shinh-Wen Hsiao, Jyh-Rhong Chou, 204, *A creativity-based design process for innovative product design*, Industrial Ergonomics, Vol. 23, 2004, pp 421-443

Tassoul,M., 2009, *Creative Facilitation*, VVSD, ISBN 047 1943517 1943518, Delft

Ulrich, K.T., Eppinger, S.D., 2004, *Product Design and Development,* McGrawHill Irwin, ISBN0-07-247146-8, New York

Walsh, V., Roy, R., Bruce, M., 1988, "*Competitive by design*", Journal of Marketing Management, Vol. 4, No.2, pp.201-217

8

Designing Disruptive Innovative Systems, Products and Services: RTD Process

Caroline Hummels and Joep Frens
Department of Industrial Design, Eindhoven University of Technology
The Netherlands

1. Introduction

There are well over hundred design processes described in literature, so why invent a new one? Over the last decade we have observed a need in our department for a process that emphasises different values than most current processes highlight. To start, we have seen a desire for a process that supports design-driven innovation, that is, we step away from incremental innovation in favour of disruptive innovation, in which disruptive refers to the absence of a well-established frame of reference for users or the market. Not only the product as such is new, but it also enables the creation of radical new meaning for the user, the market and society. We have seen a desire for design processes that can deal with this openness and complexity, in order to design open and intelligent systems that evolve during use, and which have a high level of complexity due to their adaptive, context-dependent and highly dynamic character. Next to this, the role of the designer is changing. More and more we see open platforms and design projects in which a variety of people and experts create products. We believe this has implications for the design processes used. Finally, we have seen the desire for a design process that fits self-directed learning instead of teacher-directed learning, which corresponds with educational theories like social constructivist learning.

Based on these observations on the changing face of design we present the Reflective Transformative Design process (RTD process). It is a design process, particularly aimed to support the design of disruptive innovative and/or intelligent systems, products and services, that emphasises values like openness, context- and person dependency, envisioning a new society, intuition, craftsmanship and development through reflection.

In this chapter, firstly, we elaborate on the changing field of the Industrial Design and the implications this has for design processes. Subsequently, we explain the rationale behind influential paradigms of design methodology and a variety of design processes, and show why they do not match the abovementioned changes and needs. Thereupon we introduce the Reflective Transformative Design process (RTD process) in detail. We explain how it works and elaborate on the rationale behind the model. We present the design processes of two projects, Other Brother and Ennea, to elucidate and discuss the possibilities of the RTD process to design disruptive innovative systems. We conclude the chapter by demarcating the position of the RTD process in comparison to existing processes and by explaining our plans for further development of the RTD process.

2. The changing field of Industrial Design

The field of Industrial Design is changing. At least that is the message when going to conferences like TED, the World Design Forum (WDF), ICSID World Design Congress and CHI. According to Stefano Marzano, CEO of Philips Design, we are moving towards an intellectual new renaissance based on humanistic values. Designers are catalysts for change and raise large societal questions. They are creating a vision in the first place and concrete ideas in the second (Marzano's presentation at WDF '10). Consequently the scope of design is changing. It is expanding towards all kind of systems: education, health-care, economic growth, transportation, defence, and political representation. Moreover, the role of designers is changing. Designers are dealing with a creative society in which we are all producers and consumers of value (Nussbaum, 2008). These changes have implications for educating future designers who can anticipate this changing design profession and their envisioned role, and who can even enhance these changes.

In this chapter, we describe four developments in the field of industrial design and design education, which in our opinion ask for a new view on design processes: disruptive innovation for societal transformation, intelligent systems, open design and self-directed competency-centred learning.

2.1 Disruptive innovation for societal transformation

The first development that asks for a new view on design processes is disruptive innovation, in which disruptive refers to the absence of a well-established frame of reference. Not only the product as such is new, but it also enables the creation of radical new meaning for the user, the market and society. Especially nowadays, when technology is so rapidly and innovatively created by the technology providers of the world, and at the same time when we are facing large societal challenges like healthy living and aging, sustainability and mobility, there is a need for a new type of innovation that can transform the lives of people, the way they experience and act in the world, and consequently transform society.

All designed artefacts, be it systems, products or related services, are inextricably intertwined with society; they will have a social impact as soon as they enter society. Products arise in a social context, and consequently, are a reflection of that society. Moreover, a product is a vehicle to steer society implicitly as well as explicitly; it influences the behaviour and experiences of users (Hummels, 2000; Verbeek, 2006). For example, open office layout and furnishing, which originated in the 1920s, enabled the ideas of scientific management, such as efficiency, introduced by Frederick Taylor (Forty, 1986). Designing for disruptive innovation is a way of explicitly steering society, a way of actively exploring the possibilities for social and societal transformation.

At the department of Industrial Design we are employing and researching disruptive innovation for societal transformation and we are not alone in this focus. In 2004 the British Design Council set up RED, a 'do tank' that uses transformation design to tackle social and economic issues (Burns et al., 2006). Robert Fabricant, Vice President of Creative at Frog Design, sees a shift towards 'design with intent' that has an immediate impact on user behaviour through direct social engagement (Fabricant, 2009). Bruce Nussbaum, editor of the innovation and design coverage of Business Week, states that transformation takes the best of design thinking and innovation, and integrates them into a strategic guide for the unknowable and uncertain years ahead (Nussbaum, 2008). Also Roberto Verganti (2009), professor of Management of Innovation at Politecnico di Milano, shows that we are moving towards disruptive innovation, or to put it in his words, we are moving towards design-driven innovation, that is based on a strong vision to create new markets. This type of innovation is not obtained by scrutinising user needs, which generally leads to incremental development, but by developing a strong vision that can guide disruptive innovation.

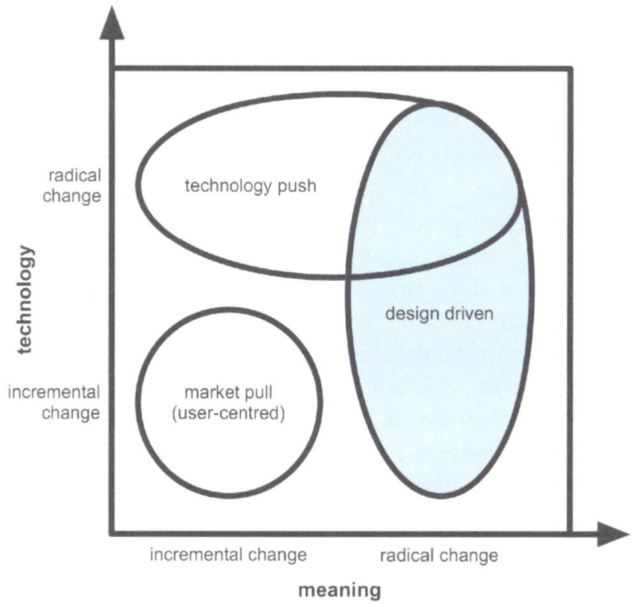

Fig. 1. The strategy of design-driven innovation through a radical change of meaning

Verganti shows that in order to realise such leaps, industry must build upon so-called interpreters, i.e. "the community of players - from artists to technology suppliers to design schools - that surround every product and deeply understand and influence how people give meaning to things." (cover, Verganti 2009). So, design schools are challenged to educate students as interpreters, with visionary design skills who can advance disruptive innovation in order to enable societal transformation. They are challenged to educate designers who are able to apply new technologies in ways that are new and daring, driven by a design vision of how our world could be, and validated by solid user research. So, educate designers who are able to transform our world, preferably in beautiful ways, instead of solving problems.

Designing disruptive innovations and envisioning societal transformation, is based on the concept of meaning, as is also indicated by Verganti. But what do we define as meaning? We adopt the phenomenological perspective in which meaning arises in interaction: "*How we think about the world is ... rooted in how we interact with it before we think, and so our intellectual thoughts cannot be used to explain away that pre-reflective experience. We move about the world, make use of the objects in it, respond to situations emotionally, act in order to change it, and so on. All these and other ways of interacting with the world give rise to its meaningfulness, so that the meaning of things in a sense, exist neither 'inside' our minds nor in the world itself, but in the space between us and the world, in the interaction*" (Matthews, 2006, p.33).

The core of phenomenology as Merleau-Ponty (2002) describes it is '*être au monde*', which means not only being in the world but also belonging to it, having a relationship with it, interacting with it and perceiving it in all dimensions. We perceive the world in terms of what we can do with it, and by physically interacting with it we access and express this meaning. Moreover, we do not perceive ourselves as one more object in the world; we perceive ourselves as the point of view from which we perceive other objects. Disruptive innovation extends our current interaction with the world and its consequential meaning for us in ways that are new to us.

What is important for designers to realise is that arising of meaning occurs during the design process too. And because designers perceive themselves as the point of view from which they perceive systems and products, they are a part of their designs. They are designing from a first person perspective while intermittently taking a third person perspective. Therefore, their designs will be meaningful for them in a different way than for someone else (Trotto et al., 2011).

Designing disruptive innovative products provides challenges for designers and industry. Designing products that do not have a well-established frame of reference for users, the market and society, requires a different design process than is often used up till now. Both Roberto Verganti (2009) and Donald Norman (2010) indicate that the classic form of human/user-centred design is not suitable for designing large radical transformations. Since users have no frame of reference, it is not possible to ask them using traditional market and user research techniques, for their needs for and requirements of these future products. The actual added value of these products becomes only clear after a certain amount of time in which users have created in interaction meaning and added value of (the services provided by) these new products (Gent van et al., 2011). Therefore, the design process needs to stimulate the development of experienceable prototypes throughout the process, also at the early start, that enable potential users to create meaning in interaction.

A second consequence of taking a phenomenological approach towards disruptive innovation and a radical shift of meaning is the importance of intuition and design action during the design process. Making enables designers to explore the unknown by trusting their senses, exploring resistance and ambiguity, and by tapping into their intuition. Dijksterhuis & Nordgren (2006) show that intuition, or unconscious thought as they call it, is better suited for dealing with complex matters than conscious thought. Designing, which is based on creating, is the highest form of (cognitive) complexity according to the Revised Bloom's Taxonomy (Anderson, & Krathwohl, 2001). Intuition begins with the sense that what is not yet could be. Intuition is necessary to make leaps. It is "*an imaginative experience ... that guides us towards what we sense is an unknown reality latent with possibility* " (Sennet, 2008, p. 213). Therefore, a design process should enable or even stimulate intuition when designing systems and products that aim at a radical shift of meaning (Trotto et al., 2011).

Concluding, a design process for disruptive innovation needs on the one hand, to enable the designer to use her intuition and design action to envision new opportunities for social and societal transformation. It should stimulate the development of experienceable prototypes throughout the entire process, also at the early start. Coming from a first person perspective, the designer can also bring in her own value system and invite certain behaviour. This envisioning and making part should be intertwined with, on the other hand, a rooting in the real life context of use, through close cooperation with all stakeholders. This way designers can explore, discover, study, anticipate and react to meaning that emerges when people interact with these experienceable prototypes and (preliminary) products in a real life context, throughout the entire design process and beyond. This way the envisioned and emerging meaning of the design-to-be can be addressed in all phases of the design process.

2.2 Intelligent systems

The second development that asks for a new view on design processes is the shift towards designing intelligent systems. When looking at the field of Industrial Design, we see that during the last decades design has shifted its focus from one person - one product (technology) interaction, to several persons via a product interaction, and it is now shifting towards a network of interactions between people and intelligent products within the context of use. Moreover, it is shifting from designing static worlds in which users adapt to objects, to co-constructed adaptive worlds in which objects and persons adapt to each other and co-evolve (Evenson et al., 2010)

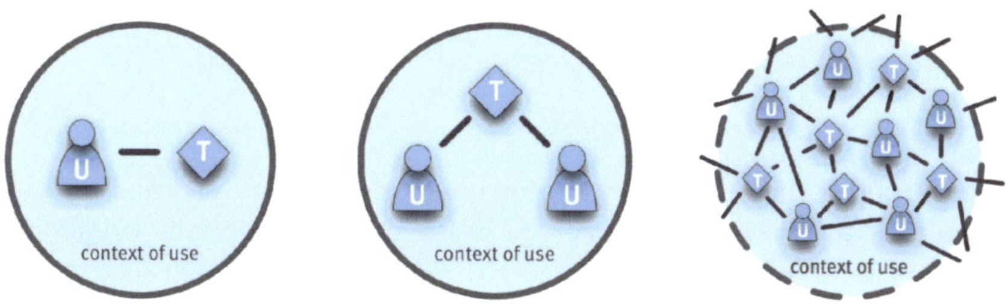

Fig. 2. Moving towards networks of interaction.

Instead of designing "closed" products and human-product interaction, a growing number of designers are moving towards developing open intelligent systems that are not finished when they leave the factory, but evolve in interaction through, for example, services and adaptation. Given the inextricably intertwinement of the designed world and people, (see previous section) not only do intelligent systems have the ability to adapt to users, but also users will adapt themselves to these systems. For example, a smart phone will adapt to its user through the different applications, personal ring-tones, background images, content and physical appearance. Newer versions move towards adaption during use, by analysing user behaviour and consequently adjusting functionality and actions accordingly. However, adaptation also works the other way around; people adapt to these systems. With the introduction of the smart phone, people have to possibility to be online and accessible 24x7 and have Internet access independent of their physical location. This has huge consequences for our behaviour, our perception of time, and our perception of leisure and work.

As a consequence of this mutual adaptation, mass customisation can grow to the level of individual user/product (system) combinations while in the mean time often unpredicted, usage patterns may emerge. Therefore designers of disruptive, intelligent systems, need a fast and good insight to what is happening with their experienceable prototypes and products in an, often increasingly diverse, social context and market (Gent van et al., 2011).

Finally, when designing intelligent systems, the complexity increases significantly. Moving towards such complexity implies that the challenges cannot be formulated exhaustively and that both challenges and solutions are not simply false or true; challenges are unique and there are multiple opportunities for solution spaces (Rittel, 1972). Consequently, designing complex systems cannot be tackled through problem solving in a linear controlled process. Kelly (1994) states that they only way to develop or manage complex systems is by letting go of control and enable the system to evolve without central authority or imposed control mechanism. Although this last take might be too bold for many designers and even goes against the grain of that what design has been until now, it stresses the need for an open process that supports evolving systems. According to Nelson (1994) it requires new strategies of design, intervention and management. He states that being undisciplined using system thinking and being out-of-control during a creative, hands-on design process are essential for creating a complex unnatural (designed) world.

Concluding, a design process for intelligent or complex systems cannot be based on linear problem solving, but needs to support openness and letting go of control. It needs to support design action and quick iterations within the real life context that give a fast and good insight to what is happening within interaction with experienceable prototypes. It needs to support the emergence of new meaning and usage patterns, preferably over a longer period of time, and support co-evolvement. Moreover, it should be possible to apply the process to an infinite number of individual user/product (system) combinations which all may bring their own unpredicted usage pattern.

2.3 Open design

The third development that has in our opinion consequences for the design processes refers to the stakeholders and participants involved in the process, especially collaborative forms like open design, co-creation and participatory design. We see open design as a specific approach to design, in which a group of intrinsically motivated people from various backgrounds develop design opportunities and solutions together in an open community, based on respect for each other's skills and expertise. Open design requires a flexible and open platform that assumes open access, sharing, active participation, responsibility, commitment to do good work for its own sake, respect, change, learning and ever evolving knowledge and skills (Hummels, 2011).

Not only designers are participating in open design; in principle everyone can participate. Key aspect is that everyone brings in their own expertise, and respects and builds on the expertise of others. Consequently, open design implies that the boundary between designers and users / customers is blurring at least with respect to motivation, initiative and needs. We agree with Bruce Sterling (2005) this does not imply that everyone is now a designer, as IKEA and many others are implying. The design profession is still something that requires many years of education and practice, like any other profession. However, (potential) users/customers now bring in their own experience as well as their specific competencies. This can be the case during the design process, for example, during co-creation sessions (Sanders, 2009) or through co-reflection (Tomico Plasencia, 2009). Especially when moving towards interactive and intelligent systems, products and services, the role of the user increases when personalising and adapting products, as was discussed in the previous section 'Intelligent systems'. For example, F# and Visual Studio on a Windows 7 mobile phone enable users to customise the functionality on their phone completely.

Given this changing role of non-designers in the design process, it is important that designers are able to co-operate with experts and users/customers, respect their competencies and simultaneously reflect on their own. More importantly, we believe that in an open design process designers should not merely function as facilitators that run co-design sessions. We believe that open design has enhanced the opportunity for and discussion about designers as subjective participants of a design process in which they are part of the solution space. Obviously, they are part of the solution space when they see themselves as potential user and customers. But based on phenomenology, designers are always an inherent part of their designs and they could even exploit that. Due to the nature of their profession they regularly take a first person perspective, as was discussed in section 2.1.

Concluding, a design process for open design should stimulate the generation of experienceable solutions to explore, generate and validate ideas and steer further developments. It should facilitate the communication between the different people/experts involved. Moreover, 'design making' opens up new solution spaces that go beyond imagination, which becomes especially important in group-settings and for designing disruptive innovative products. Especially for an open design setting we would recommend the adage: reaching quality through making quantity, which asks for a highly iterative process of generating dozens of solutions and testing them in-situ. Moreover, the process should enhance a first person perspective. Finally, open design requires a flexible and open design process that stimulates sharing between and learning from a variety of people.

2.4 Self-directed and competency-centred learning

The fourth and last important development we see that asks for a new design process is a new learning paradigm based on self-directed and competency-centred learning. Not only the focus and the way of designing is changing, as we have discussed in the previous sections, also the education of design is changing, which requires a new view on design processes, as we show in this section.

The new learning paradigm, self-directed and competency-centred learning, stems from a new view on science. Prigogine and Stengers (1984) show that the history of western thinking can be divided into three paradigms: 1) the classical-Christian view developed by e.g. Aristotle, Ptolemy and Thomas Aquinas, 2) the classical-scientific view developed by e.g. Newton and 3) quantum physics, relativity, dissipative & self-organising structure view developed by e.g. Einstein, Bohr and Prigogine (Doll, 1986).

Einstein's theory of relativity dismantled the notion of objectivity and predictability as initiated by the classical-scientific view of Newton. Where Newton's world is essentially simple and closed: it can be modelled through time-reversible laws and all complexes can be reduced to simples, Prigogine's reality is multiple, temporal and complex. It is open and admissible to change. The non-linear nature of the interconnections within complex systems, implies that such systems cannot be reduced because the information is not comprised of separate elements but distributed in a pattern of connections (Fleener, 2005). These systems do not only refer to a sub-atomic level, but to all systems from micro- to macroscopic.

As a consequence, the open, interconnected and complex character of the third paradigm disrupts the quest for certainty, truth, simplicity and objective knowledge, as is often aimed for in 'classical' research (Fleener, 2005), and in 'classical' education. In the classical teacher-centred educational approach, teachers take a third person perspective, which Doll calls the God's-eye view. They determine what the student should know and they make use of a measured and uniform curriculum, with tests that are considered objective and predictive. The new paradigm however asks for a learner-centred approach based on a transformative curriculum that emphasises and supports a variety of procedures and interpretations, depending on the learner (Doll, 1986). It asks for new perspectives and learning theories that focus on learner-world relations (Birenbaum, 2003; Segers et al., 2003). In this new paradigm, novice designers learn to learn (what, how and why) and teachers facilitate their learning. Moreover, teachers have become learners too. This will switch their role towards teaching from a first person perspective instead of a third person perspective.

Learning theories based on this new paradigm such as constructivism is gaining interest. The individual or cognitive variants of constructivism assume the locus of knowledge

construction to be in the individual learner; the social or situative variants assume this locus to be in socially organised networks (Birenbaum, 2003). Common to both perspectives, however, is the notion of activity: it is the learner who creates meaning, affected by and reflecting her socio-cultural environment. It is about learning and performing through practical application, while simultaneously acquiring theoretical skills and building knowledge. It uses the making skills of the designer as well as her analytical skills to gain knowlegde (Hummels & Vinke, 2009). It is a unity of theory and practice, where experience plays a crucial role (Dewey 1938).
We call the conceptual learning model that fits the above, self-directed and competency-centred learning, which is based on the learning model from Voorhees (2001).

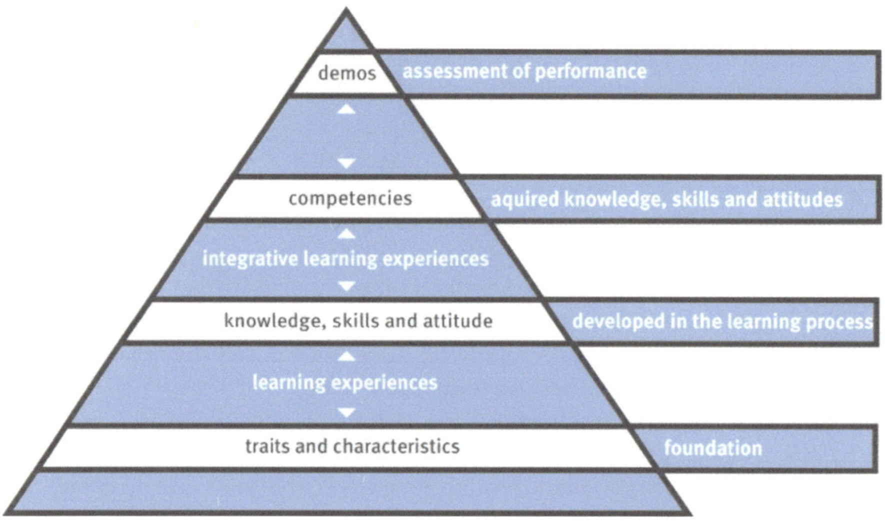

Fig. 3. Self-directed competency-centred learning model (Hummels & Vinke, 2009)

It starts from the traits and characteristics of the individual learner, who learns through doing and from experiences and thus develops knowledge, skills and attitudes in a specific context. When integrating these learning experiences the learner develops competencies which he can demonstrate when applying them. Competency-centred learning is experiential (learn by doing), exemplary (learn from specific situations), context-related (learn within a variety of contexts), reflective (in, on and for action) and it is self-directed, because it is the learner who creates meaning, which can lead to competency development.
Competency development within this paradigm and learning model follows an equilibrium - disequilibrium - re-equilibrium pattern (Piaget, 1971), where one goes from one stable state to another, in which the disequilibrium is often chaos through which one reaches order. Disequilibrium is the driving force of changing behaviour and development. Reflection and action are essential elements to regain order because they can change personal structures and ways of looking at the world and dealing with it (Doll, 1986). This fits Schön's reflective practice that is based on the ability of professionals to know, reflect and learn in and on action; to learn by doing, and through reflection gain an understanding that arises from experience (Schön, 1983). Consequently, designers need to trust their intuition, use their common sense, and dare to make mistakes, or as Schön states it, by entering into an experience, without judgment, responding to surprises through reflection,

we can learn from our actions. Or as Merleau-Ponty (2002) states, perception, through action, precedes cognition: reflection is a consequence of action.

It is important for novice designers to develop the ability to reflect in and on action as well as reflection for action, not only for designing itself, but also to stimulate learning and direct development. Especially in a learner-centred paradigm designers need to be able to direct their learning thus becoming autonomous and lifelong learners (Vinke & Hummels, 2010).

The need to become self-directed lifelong learners is becoming even more important nowadays, because the advances in science and technology follow each other so quickly that large amounts of knowledge and information get outdated rapidly. When looking at the design education of the first author (between 1985 and 1993), there has been an enormous change in focus, tools and techniques. The focus back then was mainly on 'one person - one product' interactions within a fairly static world. Most of the engineering part of the curriculum was based on mechanical engineering and hardly on digital electronics or informatics. Computers were mainly large boxes on desktops and not the mini-processors that are in all interactive devices nowadays. Consequently, functioning effectively in a rapidly changing society requires the ability to learn continuously.

So, what are the implications of a self-directed and competency-centred learning paradigm for design processes? Competency-based learning is a highly person- and context-dependent process. Since designers have different traits, characteristics and competencies, the new design process should accommodate these differences, breath flexibility and make designers aware that there are different ways to run a design project. The design process should enable chaos and a disequilibrium, next to a (re-)equilibrium, instead of breathing an atmosphere of control and rigidness. It should express that designers can make mistakes, and more importantly, that they can trust their intuition. Moreover, the design process should value design making (synthesising and concretising) next to design thinking (analysing and abstracting), and put a high emphasis on design action and experience. Moreover, the design process should emphasise and support reflection in, on and for action, not only to develop (tacit) knowledge and make decisions during the design process, but also to support novice designers becoming aware of what they have learned, and stimulate their overall development as a designer.

2.5 Implications of the changing field of design and education for the design process

In the previous sections we elucidated four developments in the field of industrial design and design education, and have sketched the implications for the design process. We do realise that a design process is merely a model of reality that emphasises certain values and downplays or even ignores other values. Professional designers have often internalised the design process, using and adjusting it based on the situation at hand. Therefore we believe that design processes are especially beneficial for students who are learning to become a designer. Novice designers have not internalised a process yet, and subsequently have not experienced and decided upon their preferred values. Based on the aforementioned developments, we are looking for a design process that makes novice designers aware of values like openness, diversity, flexibility and craftsmanship.

Next to stressing certain values, design processes are also a means to make ones activities explicit and thus have an opportunity to reflect on those actions. In addition, by making it explicit it can also smoothen the conversation to other stakeholders involved in the process, being it fellow students, participating experts and colleagues, clients or a coach. In addition, it can guide novice designers towards new possible activities within the design process.

We summarize briefly our findings from this chapter for the design process based on the aforementioned developments: disruptive innovation for societal transformation, intelligent systems, open design and self-directed competency-centred learning.
A design process for *disruptive innovation* needs to enable the designer to use her *intuition, design action* and *experience* to *envision* new *opportunities* for social and societal *transformation.* It should stimulate design *making* (synthesising and concretising) to open up new solution spaces that go beyond *imagination,* next to design *thinking* (analysing and abstracting). It should be a process that values knowledge, skills and attitudes. Moreover, it needs to stimulate *quick iterations,* reaching quality through *quantity,* by *exploring, validating* and *launching* designs in the *real life context.* By developing *experienceable prototypes* throughout the entire process, the designer gets a fast and good insight to what is happening *within interaction* in a *diverse social context* and *market.* In addition, the process needs to express and enable ingredients like *sharing, openness, uncertainty, subjectivity* and *complexity.* The design process should enable *chaos* and a *disequilibrium,* next to a *(re)equilibrium.* It should be able to deal with an infinite number of *individual user/product (system) combinations.* The process should be *flexible* and highly *person- and context-dependent;* support the *diversity* of designers to find *their preferred way* of creating design solutions within a certain *context.* The design process should preferably stimulate the awareness that designing regularly takes a *first person perspective.* Moreover, the design process should emphasise and support *reflection in, on and for action,* not only to develop *(tacit) knowledge* and *make decisions* during the design process, but also to support *novice designers* becoming aware of what they have *learned,* and stimulate and direct their *overall development* as a designer.

3. Current paradigms of design methodology

In the previous chapter we have formulated the characteristics of a process for designing disruptive innovative systems, products and services. Before developing our own process, we first explored if such a process already existed.
When looking at Dubberly's overview of design models (Dubberly, 2005), which is not an exhaustive but certainly a large collection of design models (well over eighty models) we see that a vast majority of the dozens of presented models start with some form of 'thinking' activity before moving towards synthesis, such as analysing, establishing needs, gathering and ordering information, understanding the context of use, establishing goals, planning, setting requirements, formulating the boundaries and the overall problem. Moreover, a vast majority of the presented models have a clear order and timeline line, be it linear, circular, a waterfall or wave-shaped, either with or without iterative loops and the possibility to redo certain steps.
The values that these models express seem to be incongruous with values like openness, flexibility and being out-of-control. Moreover, they seem to put cognition first, in contrast to Merleau-Ponty's stance that perception, through action, and pre-reflective experience precedes cognition: reflection is a consequence of action.
When looking at Dorst (1997) we see a similar pattern. He compares two influential paradigms of design methodology, one in which design is seen as a rational problem solving process (Simon, 1969; Roozenburg & Eekels, 1991), and one that regards design as an activity involving reflective practice (Schön, 1983).

3.1 Rational problem solving

This approach, which was introduced by Simon (1969), can be described as '… the search for a solution through the vast maze of possibilities (within the problem space) … Successful problem solving involves searching the maze selectively and reducing it to manageable solutions.' (Simon, 1969). In order to find these solutions, the designer goes through the following basic design cycle:

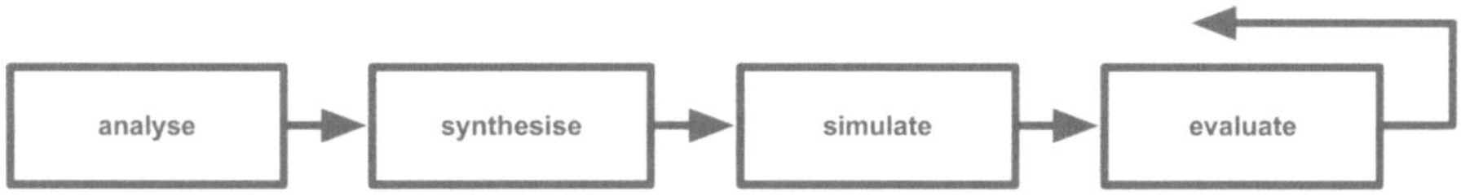

Fig. 4. Rational problem solving process.

There are many related processes that are based on this process such as the model of human-centred design activities as specified in ISO standard 13407 (Markopoulos et al., 2008). This model has comparable phases, although they are clustered differently and they put a large emphasis on participation of users (see Figure 5).

Although most of these models have iterative loops and the possibility to redo certain steps, the overall process has a clear order and timeline incorporated. When contrasting it with our conclusion in chapter two, we see several discrepancies. Firstly, this design process starts with analysis before moving towards synthesis, thus blocking approaches that start with design making and experience. In this sense, it does not give equal weight to knowledge, skills and attitudes. Secondly, although the process is iterative (the designer makes several loops of these four steps), the process is sequential and fixed. It doesn't allow for flexibility, personal freedom and context-dependency. Thirdly, although Simon stressed the ill defined and unstructured character of the design task, which we also consider important, he starts with a confined problem space, which does not comply with disruptive innovation and our search for transformation. According to the rational problem solving process, a designer can know beforehand, the width and breadth of his design challenge and its solution domain.

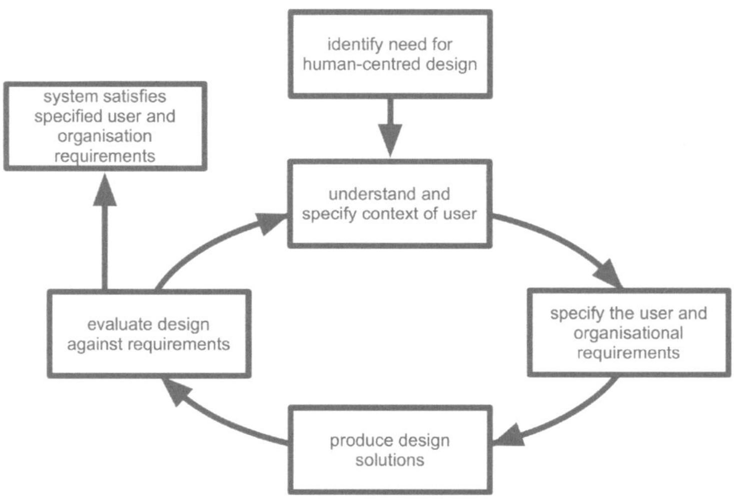

Fig. 5. Human-centred design process as specified in ISO standard 13407

We have shown that when designing complex systems, the challenges cannot be formulated exhaustively, challenges are unique and there are multiple opportunities for solution spaces.

Consequently, designing complex systems cannot be tackled through problem solving in a linear or cyclical controlled process, like the ones described above.
On the positive side, one can say that the process incorporates natural moments of reflection. For example, in the beginning of the synthesis phase, the designer is stimulated to diverge and develop many solutions, reflect on these ideas, converge and finally work towards one solution. As can be expected, these moments of reflection are guided by the requirements set within the analysis phase, which is again too limiting for our approach.

3.2 Reflective practice

Schön introduced in 1983 the reflective practitioner to stress the importance of the training of practitioners in the profession and to link the design process and task in a concrete design situation. The implicit 'knowing-in-action' is important, but this, hard to formalise, knowledge is difficult to teach. Therefore, Schön introduced reflection-in/on-action, in order to train the 'knowing-in-action' habits. In this process the designer goes through four steps:

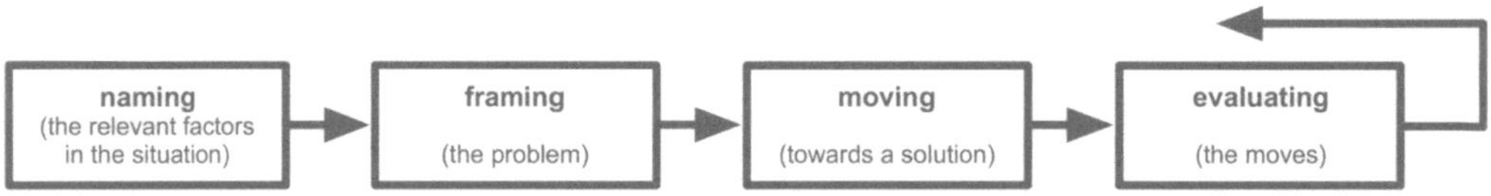

Fig. 6. Reflective practice design process.

Given the importance Schön attaches to implicit 'knowing-in-action' and reflection in and on action, the starting points of reflective practice match our conclusions in chapter two. It integrates knowledge, skills and attitude. It stimulates and acknowledges the ability of design professionals to know, reflect and learn in and on action; to learn by doing, and through reflection gain an understanding that arises from experience. Schön respects a designer's intuition, by letting her enter into an experience without judgment and respond to surprises through reflection, which is the way to learn from our actions.
So, why create a new design process and not adopt Schön's process? Firstly, the design process is rather global and it appears to offer insufficient support for our students to develop their vision and stimulate reflection. The moments of reflection are triggered by surprise during the process, which seems not enough for novice designers, because they have to develop their 'knowing-in-action' habits. Moreover, the design process is still sequential starting with naming and framing, which are both related to the analysis phases of the basic design cycle. So, in that sense, the analytical skills seem to prevail, even though Schön shows the importance of experience, making and intuition for reflective practice. Finally, although the design task is unique and context-dependent, the process as such is not flexible.

3.3 Concluding

Although there are many more design processes and approaches than we can describe here, we have concluded that most existing processes have positive and negative aspects for designing disruptive innovative intelligent systems, products and related services. Most processes use a sequential approach to gather information; a formal analysis phase precedes the creative conceptual phase. Moreover, the majority regards design action as something that implements knowledge instead as something that generates knowledge.
Since a design process is merely a model of reality which emphasises certain values and downplays or even ignores other values, we have created a new process to help novice

designers understand the principles of disruptive innovation, of designing intelligent systems, products and related services, of open design and of learning to become a designer within a competency-centred and self-directed learning setting. This process is called the Reflective Transformative Design process (RTD process) (Hummels and Frens, 2008). With our RTD process we do not aim at negating the existence and value of other used design and developmental processes. In many cases other processes can even be incorporated in the RTD process, due to its open character. Nevertheless, we want to offer a process expressing specific values for the changing field of design and design education.

4. Reflective Transformative Design process

The Reflective Transformative Design process (RTD process) is especially created to address the changing field of design and design education as discussed in chapter two. It supports designing disruptive innovative products and intelligent, open systems. Moreover, it does not only aim at supporting the creation such designs, but also aim at supporting novice designers to learn and develop while becoming a designer. In this chapter, we first explain the model, before elucidating how it supports the changing field of design and education.

4.1 The model

Developing design solutions, which are placed in the centre of this model, can be seen as a process of taking decisions based on too little information. The breadth and complexity of

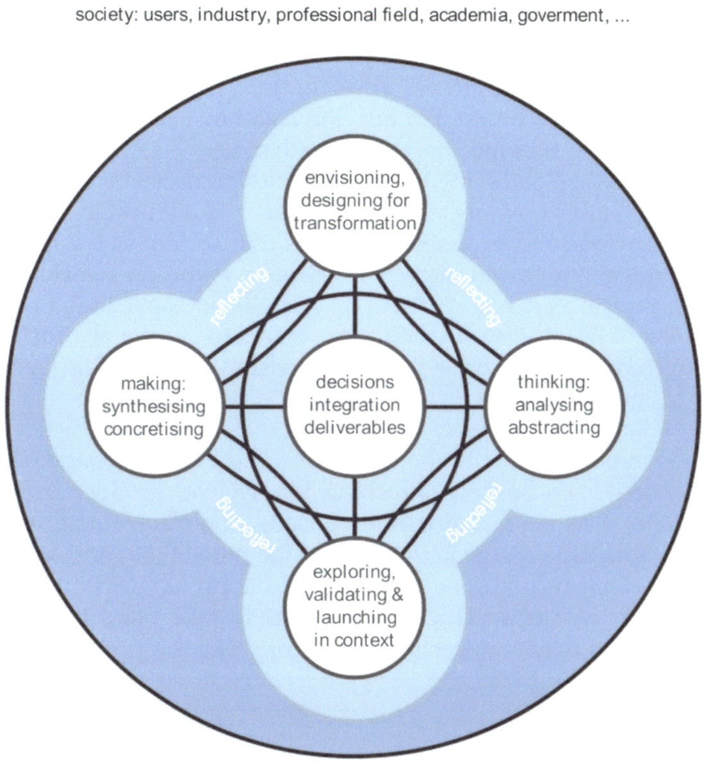

Fig. 7. The Reflective Transformative Design process.

the solution domain, and the interdependence of individual solutions, the design brief and vision make it impossible to determine beforehand if a decision is the right one. Therefore, we consider design decisions conditional. That is, a designer makes decisions to the best of her experience and knowledge. These decisions are not necessarily correct decisions, it is possible that further insight into the design challenge invalidates a decision, forcing the designer to rethink certain solutions and come up with more appropriate solutions. So decisions can change over time depending on the developments and emerging meaning. In a sense this links to characteristics like openness, uncertainty and being (partly) out-of-control. So instead of using a linear controlled process, the RTD process uses an open, explorative approach.

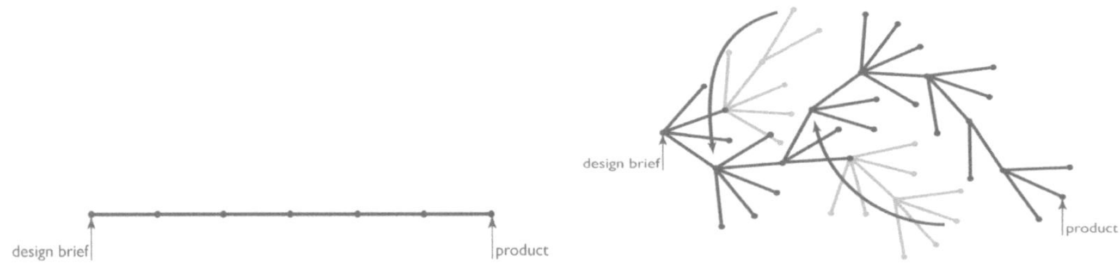

Fig. 8. A linear controlled process (left) versus an open, explorative process (right).

The RTD process knows two axes: vertically we distinguish drives and horizontally we distinguish strategies for information gathering to direct design decisions.

4.2 Drives (vertical axis)

We view the design process as a process where insight into design opportunities and the solution domain is achieved by continuous information gathering. Next to the design solution itself, we see two drives for information gathering.

The first drive is information gathering to direct the design decisions through the designer's vision (top circle). It focuses obviously on development of disruptive, innovative solutions to transform the behaviour and experience of users and society as a whole. Therefore, the RTD process encourages (novice) designers to create a vision on transformation from our current reality to a 'radical' new reality through an intelligent, open and/or complex system, product or service. This transformation can refer to personal, social and societal transformation. In the beginning of the project this vision might still be small and captured implicitly in the design brief if there is one. During the process, the vision can be developed and sharpened.

The second drive is information gathering to explore and validate design decisions in a real life context with users even beyond launching the system, product and services in the market (bottom circle). Because meaningfulness, value, and transformation are person- and context-related concepts that emerge in interaction, the possibilities and solutions have to be explored and tested extensively through experienceable prototypes and designs in the real life context. The emergence of meaning can preferably take place throughout the entire process, and later on also over a longer period of time, thus supporting co-evolvement and adaptive behaviour.

4.3 Strategies (horizontal axis)

The drives are incorporated within two strategies that generate information and that reciprocally provide focus for each other. These strategies are indicated as the basic activities that are central to academic thinking and action, consisting of analysing, synthesising, abstracting and concretising (Meijers, et al., 2005).

The first strategy revolves around design action, both synthesising and concretising, such as building experienceable prototype (left circle). Synthesising is the merging of elements into a coherent composition for a specific purpose. It goes from small to large. When concretising, one applies a general viewpoint to a specific situation or case. This action goes from large to small. This making strategy produces experiential information for the other activities in the design process. Design making enables the designer to use her intuition and through making the designer can open up new solution spaces that go beyond imagination. Reaching quality through making quantity supports decision-making.

The second strategy revolves around academic thinking: analysis and abstraction (right circle). While analysing, one unravels events, problems or systems into smaller subsets with a certain intention. So the activity goes from large to small. Abstracting does the opposite, going from small to large. It aims at making a viewpoint such as a theory, model or statement, relevant for more cases by bringing it to a higher aggregation level (Meijers et al., 2005). Academic thinking produces a more formal kind of information that (again) feeds into the connecting activities. Both strategies are valuable and should frequently alternate throughout the entire process.

4.4 Overall approach

Dependent on the person, context, or phase within the design process, designers determine where they start, and the order of the activities. This way the process supports flexibility, diversity and individuality, and it can even enhance chaos and going from a disequilibrium to a (re)equilibrium. The designers also determine how often they swap from one activity to another, although a high pace is recommended, especially during the early phases of the design process, but also during the later phases since this enables the designer to get a fast and good insight to what is happening within interaction in a diverse social context and market. As said, the RTD process is also an instrument to learn novice designers to become aware of values like openness, diversity, flexibility and craftsmanship. Moreover, the model actively supports reflection in, on and for action. The RTD process supports them to make their activities explicit and thus have an opportunity to reflect on those actions. When performing an activity within a circle, a student is stimulated to reflect in and on action, and an opportunity for reflection occurs every time the student switches activities. Therefore, we stimulate frequent changes from one activity to another, because this could help novices to

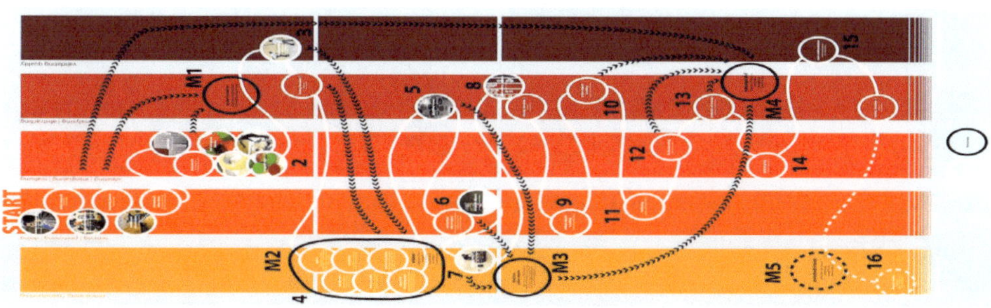

Fig. 9. A visualisation of the RTD design. The five circles were translated to rows and every activity was placed within, showing the relationship between activities.

train their reflective practice. The activity of reflection is indicated in the model (figure 7) by the lines between the mutual activities, and between the activities and the deliverables. Reflection on and for action can also be related to the entire project, learning activity or overall development. This is represented in the model by the reflection line of the outer circle.
In addition, by making the activities explicit by visualising all the steps, it can also smoothen the conversation to other stakeholders involved in the process. We encourage our design students to document their process in a schematic way (see figure 9). There is not one way for doing this; it is related to the skills of the students and their preference for a certain way of learning.

5. RTD process applied

The previous chapter described the model of the Reflective Transformative Design Process. In this chapter we describe and show the processes of two designs Other Brother and Ennea, in order to elucidate and discuss the possibilities of the RTD process to design disruptive innovative intelligent systems for transformation.

5.1 Other Brother

"The Other Brother" is a semi- autonomous device that captures images and video of spontaneous moments in the course of everyday life to enable people to re-experience these moments in a playful way. It is designed by John Helmes during his final Master's graduate project in cooperation with Microsoft Research Cambridge (UK) Lab (Helmes at al., 2009). The overall goal of this project was to design a situated, tangible object for a domestic setting, capturing natural and spontaneous social situations. It focused on more serendipitous, lightweight ways in which moments can be captured instead of conventional photo and video cameras, which require a person to take the initiative and to control the framing of the shot, leading to somewhat predictable results.
The Other Brother is the result of an iterative process using the RTD process, from initial (interactive) sketches, concepts and physical explorations towards several prototypes and a final design that were tested several times throughout the process in a home environment.

Fig. 10. The Other Brother captures images and video of spontaneous moments at home.

The project aimed at designing a disruptive innovative system. The client, Microsoft Research Cambridge (UK) Lab, did not set a specific brief up front. They gave full freedom to John as long as the project would fit the overall focus of the Socio-Digital Systems (SDS) department of MSR. After seeing a variety of projects of the SDS department, John decided to focus on connectedness in the broadest sense of the word. He started formulating the overall goal to design a family of physical objects that are intertwined and enable people to be connected again in situations they are not connected right now. In order to explore the concept of connectedness, he used an extreme user paradigm to get inspired. Therefore, he started with generating 2D and 4D sketches for re-connecting people with a social phobia, and re-connecting inexperienced Internet users to experience privacy and security.

Fig. 11. Sketches to explore social phobia (left) and to explore Internet safety (right).

After finishing several 2D and physical sketches, John reflected upon his sketches and decided to focus on the creation of an object that could capture moments spent with others in the home. Four concepts were explored using computer sketches, evaluated with MSR and one concept was selected due to its serendipitous nature.

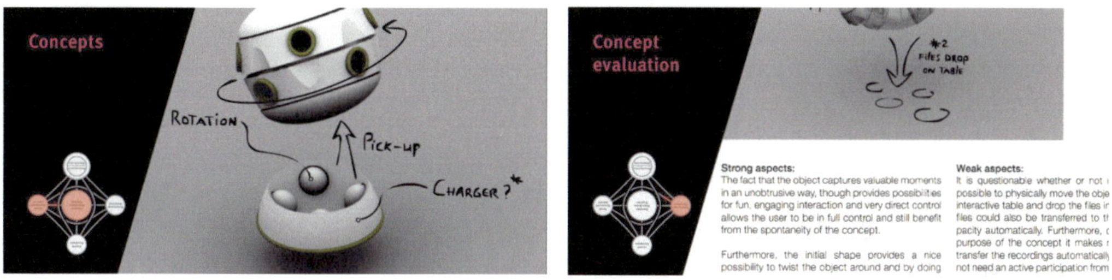

Fig. 12. Ideation and conceptualisation (left) and the evaluation of concepts (right).

Animation sketches were developed for watching the recorded files on an interactive table. Moreover, John started exploring cameras present in the market and research labs that have the option to respond in a dynamical way. This step, analysing existing products and literature, is generally done in an earlier stage when using other design processes. John explicitly decided to first develop his own vision and concepts, before being influenced by others.

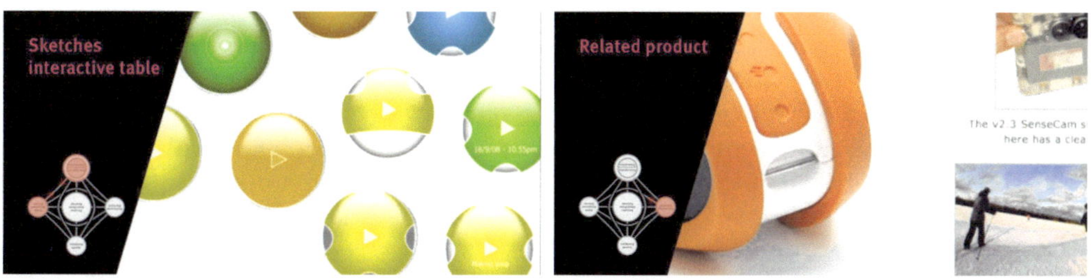

Fig. 13. Animated sketches (left) and a market evaluation of related products (right).

He sharpened his vision further by integrating a level of serendipity within the captured photos, videos and sounds. He predicted that the captured results would be much more surprising when it is not possible for the user to exactly know what is being captured. The concept was further explored by means of a first working prototype. The main component of this prototype consisted of a digital photo camera controlled by external sensors and actuators.

Fig. 14. First working prototype (left), explorative user study in the home situation (right).

Since meaning is generated in action, he initiated a first explorative user study with three families during Christmas to observe how people responded to the object and interact with it. Many questions arose during and after the study, which guided the direction of fine-tuning the vision. The fact that the first simple prototype was becoming an additional character within the group, a character taking part in the social activity, was further developed within the second prototype. The second prototype of The Other Brother was equipped with a front and back cover, a LED display and RGB-LEDs to change its colour. Moreover a website was made to view the photos made by The Other Brother, and a possible table-top interface was explored by means of a flash animation.

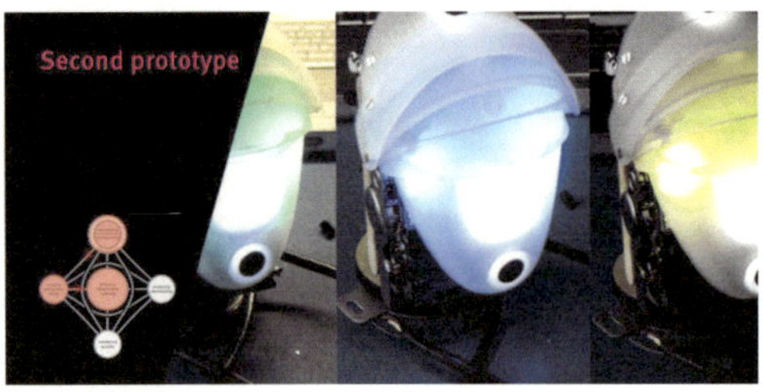

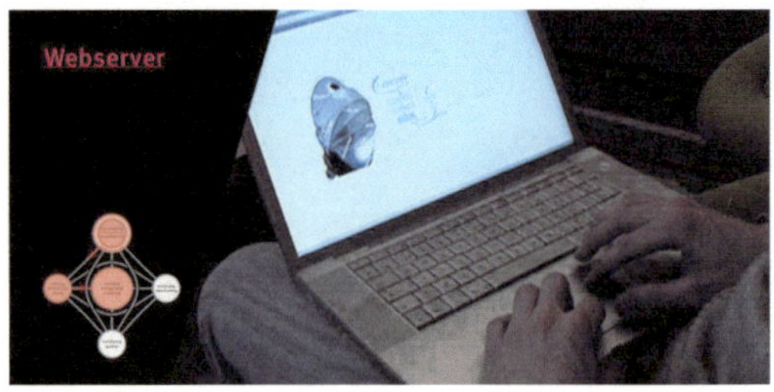

Fig. 15. Second working prototype (left) and a website to see the photos and videos (right).

After the initial deployment and re-design, a diary study in combination with an interview was executed within two different families as well as during a social meeting with a larger group of people. The main goal of this study was to deploy The Other Brother within a domestic environment for a longer period of time, and a social setting with a large group and observe people's behaviour around it. The families were instructed to use The Other Brother for one week and were asked to keep track of their activities within a provided diary. They could view the photos and videos on the web and in a photo frame. The social meeting had an open character without instructions. The outcomes of these user studies were used to adjust the vision and re-design The Other Brother.
After creating a new design using computer sketches, the iterative process continued by means of translating the design into several Solid Works structures allowing to use 3D printing technologies in order to create a museum quality model. At the end of this process John reflected on his actions, as he had done regularly during the entire process.

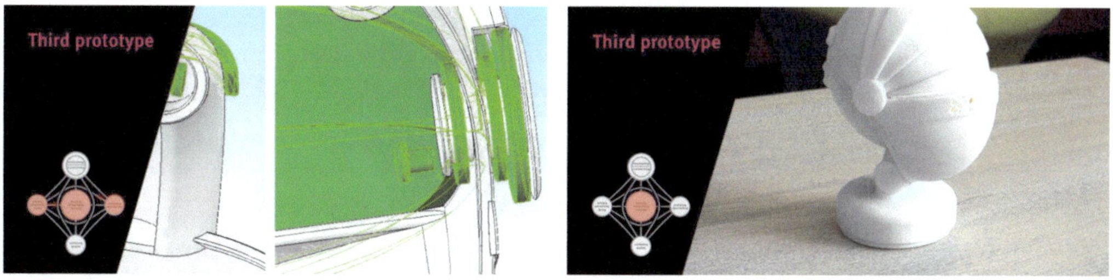

Fig. 16. Solid Works model (left) and the final working prototype (right).

It can be concluded that the Other Brother enables people to re-experience spontaneous moments from the past. Throughout the several studies it appeared that people found an emerging way of interacting with an innovative device that was radically new for them. The device and the captured fragments positively surprised them. This could have never been done without the working prototypes that John developed during the project. The studies enabled him to sharpen his vision, to study emerging behaviour and find possibilities for further development. Moreover, the RTD process enabled him to exploit his strength: ideation, conceptualisation and envisioning through interactive sketches and prototypes. John is someone who likes to use his making skills to initiate his decisions; truly reflection in, on and for action. He used this strategy at the start of the project too. The RTD process, which was new when he started using it for his project, legitimised him to start from making and envisioning, instead of analysing a problem. There were several people that questioned the validity and usefulness of starting with making and envisioning instead of analysing. But John proved them wrong. After having used the process during this project, he stuck with it and is still using it in his job.

The client, Socio-Digital Systems (SDS) department of MSR, deliberately gave him an open brief to become immersed in the concerns and questions that interested the group, and for them to be inspired by his designs. "*We were both surprised and pleased with the way this single device was being developed as part of a larger ecosystem of devices within the home. We hadn't expected such a breadth of vision.*" (feedback client). Especially that last remark fits one of the main values of the RTD process: envisioning transformation. "*By far the biggest surprise for us as a group, however, was the realisation that the device itself appeared to have a life-like quality to it.*"(feedback client). Again a remark that shows that disruptive innovation, and emerging behaviour and interaction is not something that one can imagine or reason upfront. It is something that grows in interaction, while making, envisioning, testing, analysing, creating, etcetera in a real life context during the entire design process.

5.2 Ennea

As said earlier, the RTD process is flexible, open and person-dependent, meaning that it can be used in many ways. Therefore we show the process of another project called Ennea.

Ennea is designed by Master's students Jasper Dekker, Laurens Doesborgh, Sjors Eerens, Jabe-Piter Faber an Jan Gillesen. It is a system that consists of several networked products that are coupled through an online platform. It is aimed at high school freshmen and is meant to guide them through their first year by analyzing their social behaviour and giving teachers the opportunity to aid in undesired situations (e.g. social isolation).

Fig. 17. Ennea is designed to support social behaviour of high school freshmen.

The Ennea 'nodes' are handed out to high school freshman when they enter high school and are carried by the pupils at all times. The nodes measure the proximity of other nodes and thus map the social structures in the group of high school freshmen. To be more specific, the measurements are condensed into two variables: (1) duration of contacts and (2) diversity of contacts. Based on these two variables each pupil gets assigned a 'social' role by the system that is representing her type of social behaviour. By rubbing two nodes together the pupils can see these roles temporarily appear on the small round screen of the node and can reflect on how they are doing in their new situation. Throughout the year these roles and social patterns are also discussed by teachers so as to make sure that nobody is 'left behind' and isolated.

The Ennea system was the result of a six-week master class that was sponsored by Microsoft research (USA). The students were asked to design a product or system within the context of learning and education using the RTD process (then at its infancy). The process that led to Ennea started rather analytical. By quickly cycling through analytical activities (reading literature, studying online information) and envisioning activities the students created an understanding of what their opportunity for design was to be. When they were satisfied with their vision they went on to explore through interactive and tangible sketches how their vision could take form.

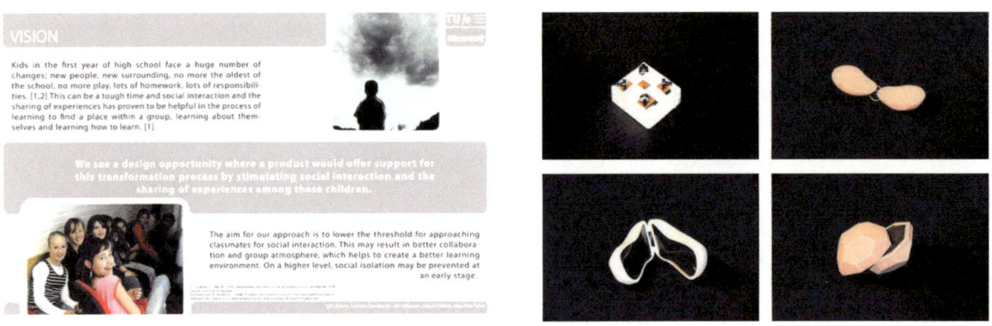

Fig. 18. Vision that led to Ennea (left) and exploration through tangible sketches (right).

They presented a vision, a scenario and a prototype that was scrutinized by coaches and fellow students. This led to a process of reflection on the starting points for their project and a rewritten vision.

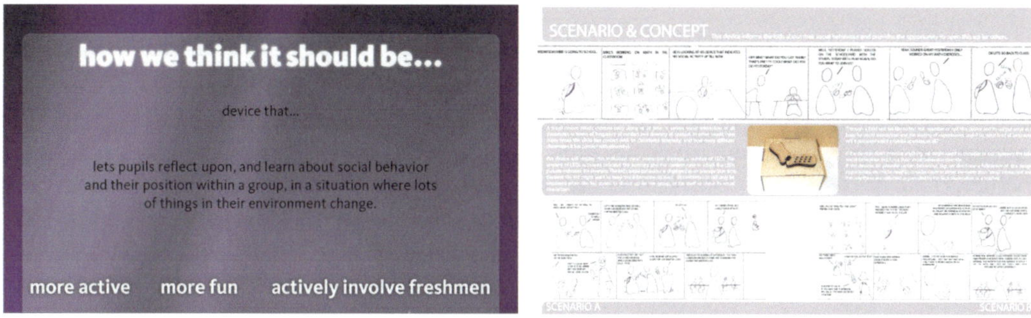

Fig. 19. Adjusted vision (left) roles temporarily appear and scenario with a first concept (right)

From there they went to the high school freshmen themselves and commenced in a co-design session in context. The results of this activity were analyzed and slowly but certainly their understanding of the design opportunity became more fine-grained. They solidified their proposal with a thorough exploration of form and interaction.

Fig. 20. Co-design session with freshmen (left), exploration of form and interaction (right).

They finalized the design by creating a set of working prototypes that were tested in context.

Fig. 21. Fully working prototypes (left) and testing designs in context with freshmen (right).

The process was very iterative in the sense that all activities were done multiple times throughout the process. Because the students reflected on what they did and on what they learned they were aware of the process they were going through (sometimes helped by coaches) and consciously steered their process through the activities of the RTD process. Therefore, the order in which the activities were done was not the same for each iteration. When looking at their process in retrospect it is striking to see that the goal of the project (that what the students wanted to accomplish) was under development almost till the moment that they started making the final prototypes. This highlights a typical characteristic of the RTD process: the process offers students the ability to keep momentum in their projects even when important decisions are still based on assumptions. The process does not fix the decision points in the process (as a sequential process does) but encourages exploration and the gathering of insight by means of different activities and reflection. The students were able to fine-tune and even change parts of their point of departure because they were filling in their assumptions while exploring and contextualizing their insights through their activities. Next to this the students commented on how the process allowed them to 'make mistakes' during the process. They were encouraged to keep up the tempo of their activities, as there was only limited time for the project. This led to quick successions of activities and many moments of 'reflection on action'. Because they started so intensively and because they made multiple, quick design cycles they found the opportunities and the dead-ends of the project early on in the process leaving them much more time to ground and fine-tune the project.

6. Conclusions and future developments

In this text we have shown that the changing field of industrial design and design education towards disruptive innovation for transformation, intelligent systems, open design and self-directed competency-centred learning, asks for a new view on design processes. The Reflective Transformative Design process is created to address these developments and emphasis values like openness, flexibility, diversity, context- and person dependency, envisioning a new society, intuition, craftsmanship, design making and design thinking, knowledge, skills and attitudes, and development through reflection. Given the importance of these values, one can regard the RTD process as an attitude rather than a method. Moreover, we have seen that the process forms the solution. For example, The Other Brother was a result of John Helmes' attention for envisioning, making and testing in a real life context and his desire to follow his intuition and get surprised in his search for disruptive innovation. We have seen similar results during the class *Multi-disciplinary perspectives on the design process?* that was run together with Panos Markopoulos. In this class we compared and discussed the different perspectives, strengths and weaknesses of three design processes in comparison with the RTD process with help from Philips Research (Value Proposition House), Bright Innovation Pittsburgh (Sales, Learn, React, Build), Astcon Rozwiazania Informatyczne (Agile: SCRUM) and Microsoft Research Cambridge (RTD process). The results from this class showed that every process stresses specific values and has specific outcomes. The RTD process appeared to be especially suitable for creating a flexible product vision for unknown needs, see figure 22.

We believe that design processes including the RTD process are especially beneficial for novice designers who are learning to become a designer. It is a means to stress certain values, to make ones actions explicit and thus have an opportunity to reflect on those actions, to smoothen the conversation to other stakeholders involved in the process, and to guide novice designers towards new possible activities within the design process. Over the years designers find their own strength and weaknesses, and preferred approach. They have incorporated the process and reflection in and on action has been internalised.

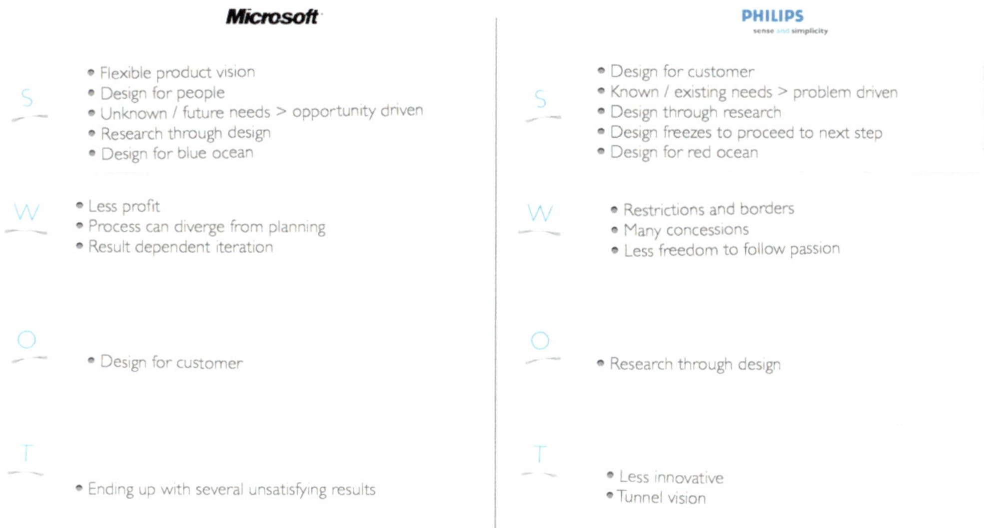

Fig. 22. SWOT analysis comparing the RTD process used at MSR Cambridge (left) and the Value Proposition House process used at Philips Research (right).

Consequently, the different activities within the RTD process will intertwine and be less discernable. The preference for certain activities within the RTD process will differ per person, resulting in a kind of personal process profile for every designer, see figure 23..

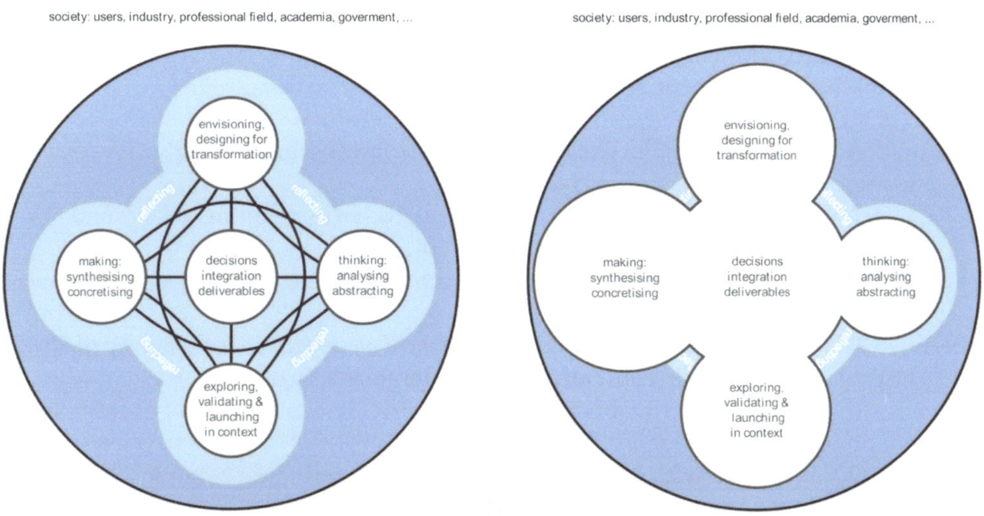

Fig. 23. Over the years the different activities within the RTD process will intertwine and result in a kind of personal process profile (right).

We have used the RTD process the last few years and it is still developing. Students experiment with a variety of visualisations to encourage reflection. Moreover, we are exploring together with industrial partners how we can extend the value for industry. Furthermore, we are exploring the need for new design methods that accompany the process. For example, PhD student Carl Megens is developing a flexible and dynamic Personas method that complements the RTD process. Finally, we are exploring the possibilities of real life settings during the entire design process. We are developing Experiential Design Landscapes to trigger and study emerging patterns in interaction with disruptive innovative systems. The data obtained from Landscapes is monitored, and via data-mining techniques emerging patterns are detected and responded upon. In this manner design synthesis, emerging behaviour and market analysis become integrated (Gent van et al., 2011).

7. Acknowledgment

We like to thank all staff members and students who helped developing the RTD process, including Oscar Tomico, John Helmes, Jasper Dekker, Laurens Doesborgh, Sjors Eerens, Jabe-Piter Faber, Jan Gillesen, Panos Markopoulos, Carl Megens, Aarnout Brombacher, Sabine van Gent and Lu Yuan.

8. References

Anderson, L. W., & Krathwohl, D. R. (2001). *A taxonomy for learning, teaching and assessing: A revision of Bloom's Taxonomy of educational objectives: Complete edition*. Longman: New York.

Birenbaum, M. (2003). New insights into learning and teaching and their implications for assessment. In M. Segers, F. Dochy and E. Cascallar (eds.), *Optimising New Modes of Assessment: In search of Qualities and Standards*. Dordrecht: Kluwer.

Burns, Cottam, Vanstone and Winhall, (2006). *Red paper 02; Transformation Design*. London, UK: Design Council. Last accessed July 26, 2009: http://www.designcouncil.info/mt/RED/transformationdesign/

Dewey,J. (1938). *Experience and education*. New York: Touchstone

Dijksterhuis, A. and Nordgren, L. (2006). A theory of unconscious thought. *Perspectives on psychological science*. 1, 2, pp. 95-109

Doll, W. (1986). Prigogine: A New Sense of Order, A New Curriculum. *Theory into Practice, Beyond the Measured Curriculum* 25(1), pp. 10-16.

Dorst, K. (1997). *Describing design: a comparison of paradigms*. PhD Thesis, Delft University of Technology, The Netherlands.

Dubberly, H. (2005). *How do you design? A compendium of models*. Last accessed April 26, 2011: http://www.dubberly.com/articles/how-do-you-design.html

Evenson, S., Rheinfrank, J. and Dubberly, H. (2010). Ability-centered design: from static to adaptive world. *Interactions*, 17(6), pp. 75-79

Fabricant, R. (2009). Design with intent. How designers can influence behavior. *Design Mind*, Issue 10. Last accessed July 26, 2009: http://designmind.frogdesign.com/articles/power/design-with-intent.html

Fleener, M.J. (2005). Introduction: chaos, complexity, curriculum, and culture. In: W. Doll, M. Fleener, D. Trueit and J. Julien (eds). *Chaos, complexity, curriculum, and culture: a conversation*. New York: Peter Lang, pp. 1-17.

Forty, A. (1986). *Objects of desire: design and society* 1750 - 1980. London: Thames & Hudson.

Gent van, S., Megens, C., Peeters, M., Hummels, C., Lu, Y. and Brombacher, A. (2011). Experiential Design Landscapes as a design tool for market research of disruptive intelligent systems. *Conference proceeding CADMC* 2011 (Cambridge, UK, September 7-8, 2011).

Helmes, J., Hummels, C. and Sellen, A. (2009). The Other Brother: Re-experiencing spontaneous moments from domestic life. *Proceedings of the 3rd international Conference on Tangible and Embedded Interaction 2009* (Cambridge, UK, February 16 - 18, 2009). TEI '09, pp. 233-240.

Hummels, C. (2011). Teaching attitudes, skills, approaches, structures and tools. In: B. van Abel, L. Evers, R. Klaassen and P. Troxler (Eds.) *Open design now; why design cannot remain exclusive*. Amsterdam: BIS publishers, pp. 162-167.

Hummels, C. (2000). *Gestural design tools: prototypes, experiments and scenarios*. Doctoral dissertation, Delft University of Technology.

Hummels, C.C.M., Frens, J.W. (2008). Designing for the Unknown: A Design Process for the Future Generation of Highly Interactive Systems and Products. *Proceedings of the 10th International Conference on Engineering and Product Design Education – EPDE2008*, Barcelona, Spain, Sep. 4-5, 2008, pp. 204-209.

Hummels, C. and Vinke, D. (2009). *Eindhoven designs; volume two: Developing the competence of designing intelligent systems*. Eindhoven University of Technology, The Netherlands.

Kelly, K. (1994). *Out of Control: the new biology of machines, social systems and the economic world*. Basic Books.

Markopoulos, P., MacFarlane, s., Hoysniemi, J. and Read, J. (2008). *Evaluating children's interactive products: principles and practices for interaction designers*. Burlington, MA: Morgan Kaufmann Publishers.

Matthews, E. (2006). *Merleau Ponty. A guide for the perplexed*, Continuum: London, UK.

Meijers, A., Overveld van, C. and Perrenet, J.C. (2005). *Criteria for academic bachelor's and master's curricula*. Eindhoven: Technische Universiteit Eindhoven

Merleau-Ponty, M. (2002). *Causeries 1948*. Seuil, Paris.

Nelson, H.G. (1994). The necessity of being 'un-disciplined' and 'out-of- control'; design action and system thinking. *Performance Improvement Quarterly* Vol. 7/ No. 3.

Norman, D. A. (2010). Technology first, needs last: the research-product gulf. *Interactions, 17*(2), pp. 38-42. http://interactions.acm.org/content/?p=1343

Nussbaum, B. (2008). "Innovation" is dead. Herald the birth of "transformation" as the key concept for 2009. *BusinessWeek; NussbaumOnDesign*. Last accessed April 26, 2011: http://www.businessweek.com/innovate/NussbaumOnDesign/archives/2008/12/innovation_is_d.html?campaign_id=rss_ blog_nussbaumondesign

Piaget, J. (1971). *Biology and knowledge*. Chicago IL: University of Chicago press.

Prigogine, I. and Stengers, I. (1984). *Order out of chaos*. New York: Bantam.

Rittel, H. (1972). On the planning crisis: systems analysis of the 'first and second generations'. *Bedrifts Okonomen*. no. 8, pp. 390-396.

Roozenburg, NFM and Eekels, J. (1995). *Product Design: Fundamentals and methods*. Wiley: Chichester

Sanders, L. and Simons, G. (2009). A social vision for value co-creation in design. *Open Source Business Resource*. December 2009: Value Co-Creation. Last accessed April 26, 2011: http://www.osbr.ca/ojs/index.php/osbr/article/view/1012/973

Schön, D. (1983). *The Reflective practitioner*. New York: Basic Books.

Segers, M., Dochy, F. and Cascallar, E. (2003).The Era of Assessment Engineering: Changing Perspectives on Teaching and Learning and the Role of New Modes of Assessment, in M. Segers, F. Dochy and E. Cascallar (eds.), *Optimising New Modes of Assessment: In search of Qualities and Standards*, Dordrecht: Kluwer Academic Publishers.

Sennett, R. 2008. *The craftsman*. Penguin Books, London.

Simon, H.A. (1969). The sciences of the artificial. MIT Press: Cambridge MA

Sterling, B. (2005). *Shaping Things*. MIT Press: Cambridge.

Tomico Plasencia, O. (2009). Co-reflection: user involvement aimed at societal transformation. *Temes de disseny, 26,* pp. 80-89

Trotto, A., Hummels, C. and Cruz Restrepo, M. (2011). Design-driven innovation: designing for points of view using intuition through skills. *Conference proceedings DPPI* (Milan, Italy, June 22-25, 2011).

Verbeek, P. P. (2006). Materializing morality - Design ethics and technological mediation. *Science, Technology and Human Values*, 31(3), pp. 361-380.

Verganti, R. (2009). *Design Driven Innovation – Changing the Rules of Competition by Radically Innovating what Things Mean*. Boston, MA: Harvard Business Press

Voorhees, R.A. (2001). Competency-Based learning Models: A Necessary Future. *New Directions for Institutional Research*, Vol. 2001, Issue 110, pp. 5-13.

Vinke, D. and Hummels, C. (2010). (2010). Authentic assessment for autonomous learning. *ConnectED 2010 – 2nd international conference on design education*, 28 June – 1 July 2010, Sydney, Australia.

9

TRIZ-Based Design of Rapid 3D Modelling Techniques with Formative Manufacturing Processes

César Cárdenas[1], Yuliana Rivera[2], Ricardo Sosa[2] and Oscar Olvera[2]
[1]*The Distributed and Adaptive Systems Lab for Learning Technologies Development/Mechatronic Department*
[2]*Innovation in Design and Manufacturing Research Chair*
Tecnológico de Monterrey – Campus Querétaro
México

1. Introduction

By accelerating the new product development process, manufacturers remain competitive (Zailani et al. 2007). Physical modelling helps in this decision making process by allowing real visualization of information about the thing the model represents (Kupka 2010). Two particular three-dimensional techniques are used in physical modelling, mock-up and prototyping. A mock-up is a scale or real-size model of a design or device, used to teach, demonstrate, evaluate, and promote among other purposes. A prototype is a physical model with the most important system functionalities implemented on it. Therefore, a prototype may be used as proof of concept for the new product. A mock-up is less expensive since it requires less material and less time to be built. Most of the mock-up techniques remain free handwork based. Some of the materials used for mock-up are clay, paper, wood, plastic, and metal. A mock-up is considered a prototype if it provides some functionality of a system and allows the test of a design. Several rapid prototyping techniques have been proposed to accelerate the new product development process (Chua et al. 2010). Rapid prototyping is defined as the automatic construction of physical objects using additive manufacturing technology. Rapid prototyping is also known as solid freeform fabrication, rapid manufacturing, layered manufacturing, additive fabrication, additive manufacturing or rapid manufacturing. Because the quality of the final product obtained by rapid prototyping, it has extended its original intend to discrete manufacturing a nd fine-art applications. The traditional process includes a computer-aided design stage that convert the three-dimensional object into two-dimensional layers, then the rapid prototyping machine builds the three-dimensional object by depositing each two-dimensional layer by means of depositing liquid, powder, or sheet material which are joined together to produce the final version of the three-dimensional object. The main advantage of additive manufacturing is the ability to create almost any shape or geometric feature (Chua et al. 2010). Most of the rapid prototyping techniques have been automated. Because most of the rapid prototyping techniques built a mock-up instead of a real prototype, we think the term has been misused. Strictly speaking, rapid mock-up should be used instead of rapid prototyping. If the new concept is in the first stages of design (i.e. the ideation stage), a designer may use a mock-up to refine the solution proposal; once the solution has been chosen, a prototype can be built to present the definite solution before manufacturing. Besides the two three-dimensional physical modelling techniques presented above, sketching is also used during the first's stages of the design process to accelerate the new product development process. Sketching is the means that architects, designers, artists and sculptors use to represent, visualize and study their concepts of three-dimensional objects. Traditionally, sketching has been done with pencils and paper, resulting in a set of two-dimensional drawings representing three-dimensional objects. The current process of design is, usually, a sequence of two-dimensional hand sketching, two-dimensional computer drafting, three-dimensional computer modelling, and finally, rendering (Hopkinson et al. 2006). In recent years, three-dimensional sketching has gained popularity as an efficient alternative to conventional three-dimensional geometric modelling for rapid prototyping; as it allows the user to intuitively generate a large range of different shapes. In this chapter, we propose a new rapid three-dimensional physical modelling technique based on the wire bending structure approach that goes beyond three-dimensional modelling and before the rendering process. Designers might need a physical model before rendering in order to refine their concept. This new rapid physical modelling manufacturing process builds a three-dimensional sketch of the concept. Since this manufacturing process is added in the early stages of design we have called it rapid three-dimensional wireframing manufacturing process or rapid three-dimensional wireframing for short (rapid 3D wireframing). As we said before, this new rapid mock-up technique is based on wire bending structures. Furthermore, to design a machine that automates this new rapid 3D wireframing manufacturing process we propose a methodology supported by TRIZ principles. TRIZ is also known as the theory of inventive problem solving and is very well known in the industry worldwide (Orloff 2010). To validate this methodology we carried out several design process experiments. First we carried out experiments on wire bending freeform fabrication, and then we performed experiments using the proposed methodology and compared results with the first experiments. Both experiments were executed by mechanical designers of sophomore and senior levels at one university. A third experiment was executed and consisted on designing machines that automate the rapid mock-up technique. We present relevant statistics and results found in these experiments. Furthermore, we dedicate a section to explain our advancements in the construction of the first prototype of our rapid three-dimensional sketching machine. Finally, we provide conclusions and future work.

2. Engineering design thinking and rapid prototyping

Engineering design is the process that engineers follow to device a new product or system. In its traditional form, it is a sequential set of activities (Pahl et al. 2007). The activities are the following: identify the need or the problem, research about the need or the problem, develop possible solution(s), select the best possible solution(s), construct a prototype, test and evaluate the solution(s), communicate the solution(s). A first generation of the product or system is finished when all the steps are completed or when a first cycle is finished. Next generations of the product or system will follow after the first iteration in the engineering design cycle. Depending on the specialization, professionals practice every step differently. There are a plenty of techniques for each step and the more adequate is related to the product or system domain (Kamrani and Nasr 2010). In a widest form, the engineering design process is embedded in the Product Lifecycle Management (PLM) philosophy (Saaksvuori & Immonen

2010). In the PLM philosophy, the steps are: imagine, define, realize, service, and dispose. Motivated by sustainability efforts, the PLM cycle has been extended from the realize step to maintain and retire steps. This new PLM is also known as closed-loop PLM (Kiritsis 2010). A methodology that integrates both the engineering design process and the PLM philosophy has been recognized as a new engineering education paradigm (Crawley et al. 2010). Furthermore, the engineering design process can be matched to the project management cycle as well (Lessard 2007). In (Cárdenas 2011), we present a match between the engineering design process, the project management process, the service-learning process at Monterrey Tech and an integrated course we teach at our university focused on designing socially relevant system for social change. The process can also be found in (Cárdenas 2009). In Table 1, we present the match between the processes we mentioned above.

Framework	Stage 1	Stage 2	Stage 3	Stage 4
Engineering Design Process	Identifying the need or the problem + Research about the need or the problem	Develop possible solution(s) + Select the best possible solution(s)	Construct a prototype + Test and evaluate the solution(s)	Communicate the solution(s)
Product Lifecycle Management	Imagine	Define	Realize	Service + Dispose / Maintain + Retire
Project Management	Defining the project	Planning the project	Executing and controlling the project	Delivering the project
Integrated Course (Cárdenas 2009)	Social problem research	Concept generation + Concept pre-evaluation + Concept documentation	Concept development + Concept documentation	Concept presentation + Concept documentation
Service-Learning at Monterrey Tech (QEP)	Social problem formulation	Solution proposal	Planning and executing proposal	Assessment of social impact + Reflection from the experience + Ability to argue and use sources of information
INNOWIZ	Problem definition	Idea generation	Idea selection	Idea communication
TRIZ	Finding a problem	Abstractize the original problem to find the general contradiction	Use the general principles to solve the general problem	Concretize the general solution to the original problem

Table 1. Match between general engineering design frameworks.

Table 1 is presented to provide a systemic view of the engineering design thinking. Additionally, we add the INNOWIZ[1] design framework. The INNOWIZ framework synthesizes the design thinking (Plattner et al. 2010). INNOWIZ creators state that any stage of the general design process can be divided in the four stages (like a fractal). At the table, we also added the general TRIZ methodology (Orloff 2010). The fact that people has embedded the engineering design process in their thinking has been recognized as engineering design thinking (Dym et al. 2006).
Now based on stages 1 and 2 from Table 1, we present in Figure 1 the innovation funnel (Buxton 2007) for our integrated course (Cárdenas 2009). The innovation funnel shows that several techniques are used to decide the final concept (or solution). The innovation funnel is composed by several divergent and convergent phases. The reader can notice that Rapid Prototyping might also be used not only after the concept is finally selected to demonstrate or prove aspects of the design but also at different moments during the innovation funnel. In such sense Rapid Prototyping can be used to help defining the concept of the system (e.g. in terms of size and form).

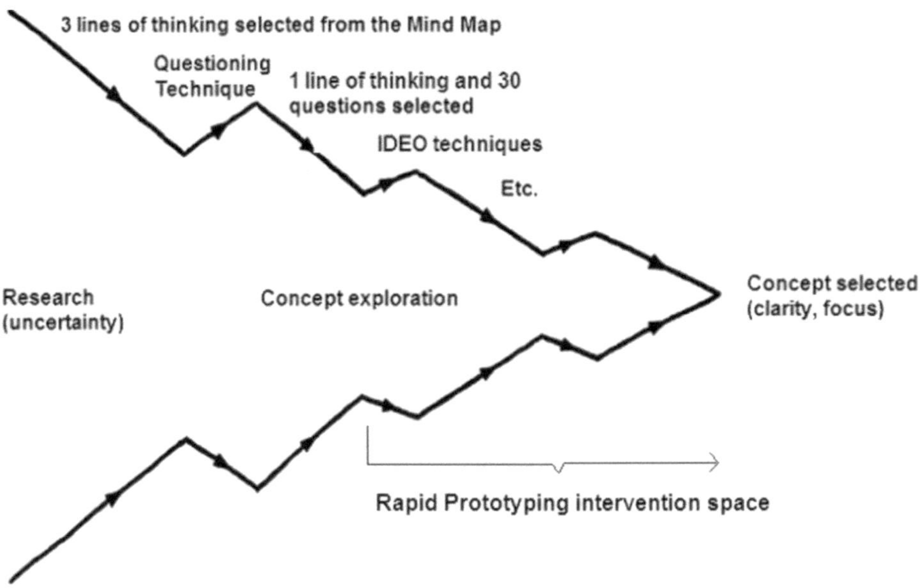

Fig. 1. Innovation funnel for our integrated course (Cárdenas 2011).

Rapid Prototyping has been traditionally used after the concept is defined and as a way to validate the concept. Several techniques for rapid prototyping have been developed (Chua et al. 2010), and most of them are based on the layered principle. Because of that, the term additive manufacturing has been recently adopted to describe rapid prototyping techniques (Gibson et al. 2009). The application of Rapid Prototyping in early stages of the design process has been previously proposed (Simondetti 1998). The time extent for applying Rapid Prototyping in the innovation funnel as indicated in Figure 1 is known as the conceptual design phase (Bruno et al. 2003).

[1] http://www.innowiz.be/

There are several ways to accelerate the new product development process. Concurrent engineering (Cha et al. 2003), Time to Market (TTM) (Smith and Reinertsen 1998), and Rapid Prototyping (Kamrani and Nasr 2010). In our previous work (Cárdenas 2009), we used the concurrent engineering approach up to the concept is defined then for rapid prototyping we use an open source hardware platform named Arduino™,[2]. For the mock-up part which comprises the conceptual design phase we are exploring new paradigms. In the following section we will explain this new paradigm

3. Formative processes and TRIZ principles for new rapid physical 3D modelling techniques

Frequently, the designer uses two-dimensional sketching and computer-aided for conceptual design (Buchal 2002) or physical modelling using the materials mentioned in section 1 (Schrage 1993). Hand-based techniques are used because their flexibility, adaptability, creativity generation, and cost (Jenkins and Martin 1993).

Traditional rapid prototyping techniques belong to additive manufacturing processes. Manufacturing processes are mainly based on subtractive manufacturing processes (Suh et al. 2010). In this chapter we exploit the less known and used family of formative manufacturing processes (Buswell 2007) to build a new rapid 3D physical modelling technique. Formative processes have been used mostly in art. There are many examples of this type of processes in sculpture[3]. They have been also used in jewels since thousands of years back. Formative manufacturing processes have the following advantages: they are environmental friendly and economic since the material can also be composed by wasted material. On the contrary, subtractive and additive manufacturing processes generate wasted material and in particular additive manufacturing processes are very expensive. Formative manufacturing processes use only the necessary material without almost any waste. The Origami technique (Demaine and O'Rourke 2007) is an example of formative manufacturing processes. Many wire bending based products are fabricated in mass. Some examples are jail birds, wire baskets, cookware tools, mice tramps, and hooks among many others.

The new rapid physical modelling technique we propose in this chapter is called Rapid Physical 3D Wireframing. The name comes from the final 3D object we want to achieve. A wire-frame model is a visual presentation of a three dimensional or physical object used in 3D computer graphics. In our case we intend to reproduce a physical 3D wireframe object. We do not want to call rapid physical 3D sketching, rapid 3D prototyping, and neither rapid physical mock-up. The reason is that the final product of this process is not a sketch, not a mock-up, and neither a prototype. There are a plenty of patents and commercial machines that blend wires but their use is not for the conceptual design phase.

TRIZ is a problem-solving approach developed from the patent mining experience of the Russian Genrich Altshuller and his colleagues. TRIZ in English means Theory of Inventive Problem Solving (TIPS) (Orloff 2010). Altshuller discovered that at least 80% of the patents were based on some general principles. Training people on such general principles gives them the possibility to invent solutions to problems in a structured form. The main TRIZ concept is contradiction. Technical contradictions emerge when two associated necessities from a product or problem are in conflict. The key issue in TRIZ methodology is to find the

---

[2] http://www.arduino.cc

[3] http://www.wirelady.com/

main contradictions in the technological innovation. According to Altshuller, there are three categories of contradictions: technical, physical and human. The methodology proposed by Altshuller consists on a series of the following sequential steps: finding a problem, abstractize the original problem to find the general contradiction, use the general principles to solve the general problem, concretize the general solution to the original problem. In Table 1 we have included this general procedure as a general engineering design thinking methodology. The use of this methodology will be explained in the following section.

4. Research contributions and questions

The contributions of this chapter are the following. First, a new rapid 3D physical modelling technique is proposed, this technique is based on formative manufacturing processes. We named this technique Rapid 3D Wireframing. The technique is expected to be used in early conceptual design phases of the new product development process. Therefore, it modifies the industrial design process but this is not evaluated in this chapter. Second, a TRIZ-based design process is proposed to reduce the complexity of the mechatronic design to print the object (Rapid 3D Wireframing). We have named MDSU this design process. Since the Rapid 3D Wireframing technique is new, so the MDSU design process and the application of TRIZ principles to the design of this kind of mechatronic systems. MDSU stands for Mesh, Unfolding (*Desdoblado*), Separation and Union (MUSU in English). The MDSU approach reduces the degree of freedom necessary in mechatronic systems to automate the process. These four sub-processes belong to TRIZ principles. It is expected that the Rapid 3D Wireframing technique will be automatic; therefore a first prototype will be explained briefly.
In this chapter the following research questions are explored: the implementation of MDSU will ease the design process of the mechatronic system; the implementation of MDSU improves the design thinking process and reduce the development time of the mechatronic system; the implementation of the Rapid 3D Wireframing technique improves the work conditions of the designers; the implementation of the Rapid 3D Wireframing technique reduces the designers competencies related to the conceptual design phase; the implementation of Rapid 3D Wireframing improves the new product design process. Through our experiments we will try to answer the previous research questions.

5. Experiments

Our main objective is to define a design process of mechatronic systems for Rapid 3D Wireframing techniques. We study two design processes executed by novice students and advanced students in mechanical engineering. The two design processes are freeform and a design process inspired in TRIZ principles.
Students groups were selected in such a way that we can observe differences between the cognitive, interpretative and creative processes. Novice students are from the first semester of mechanical engineering (sophomore). Advanced students are from the last semesters of mechanical engineering (senior). It is expected that both groups are very different in the cognitive, interpretative and creative processes.
The general statistics of the universe of students studied were the following. There were 95 students from 4 groups in total (2 novice and 2 advanced), 57 novice students and 38 advanced students. From the general information got from the students, 95% and 84% of the novice and advanced students respectively were man. 86% and 39% of novice and advanced students

respectively were between 15 to 20 years old; 14% and 42% respectively were between 21 and 25 years old; 19% of the advanced students were more than 25 years old. 70% and 58% of novice and advanced students respectively do not have industrial experience.

5.1 First experiment

Work done by hand can achieve a high level of quality. There are a lot of examples from the art craft domain. Automate such manufacturing processes is a highly complex task since it demands mechatronics systems with many degrees of freedom. To propose a design process for wirebending we must evaluate first the work done by hand. The first experiment was entitled "freeform construction of 3D objects by means of wire bending". First, the students were asked to register general information. Then, a brief introduction to the rendering process was given; the main focus was to provide a background about the wire-frame version of objects. Wireframing is used as an intermediate step to obtain rendering. Wireframing is an implemented function of several CAD/CAE platforms. Wireframing form depends on the finite element chosen (e.g. circle, square, triangle, etc.). Wireframing can be also seen as a kind of mesh. Third, students were asked to make three sketches where they will present the mesh of a cup (divergent phase). The cup was chosen because it is a well known form and a common object found everywhere. They were asked to sketch the cups with three different meshes. For the convergent phase, students were asked to choose the mesh based on some criteria: originality, design, easy to build, etc. Fourth, students were provided by enough material of 16 AWG wires (same size that will be used in the expected machine). They were asked to build their prototype by hand. Figure 2 shows two pictures of this experiment. Finally a questionnaire was applied to the participant students. In the following section we will present the more important results of this experiment.

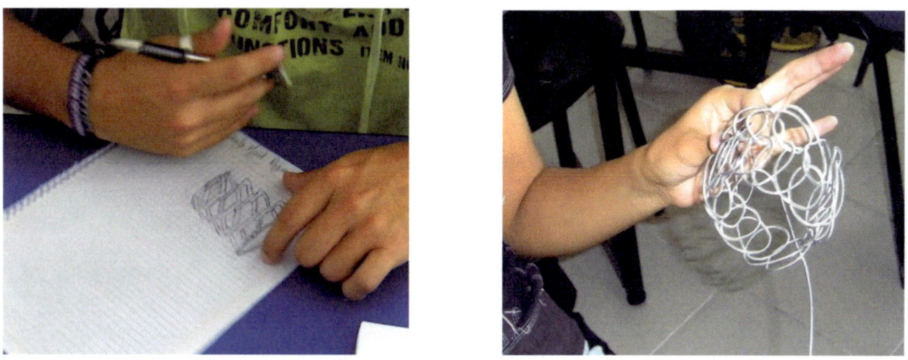

Fig. 2. Pictures from the first experiment.

5.2 Second experiment

Once the formative process was known by the students and they provided their qualitative evaluation to such process. In the second experiment we proposed a methodology that students followed also by hand. The objective of this experiment was to compare qualitatively the freeform style previously experimented with our methodology to manufacture the 3D object by wire bending. Therefore, the second experiment was entitled "MDSU-based construction of 3D objects by means of wire bending". The MDSU is our proposed methodology. It stands for Mesh, Unfolding (*Desdoblado*), Separation and Union

(MUSU in English). The MDSU approach proposed here reduces the degree of freedom necessary in mechatronic systems to automate the process. These four sub-processes belong to TRIZ principles. Mesh and Separation match with the TRIZ Segmentation principle which consists on divide the object in independent parts. Recall that mesh is compared by finite element where each element is independent. Unfolding phase match with the TRIZ Flexible Shells and Thin Films principle which suggest to use thin and flexible surfaces instead of three-dimensional structures. A simplification of this principle to wireframing takes in consideration only the vertices. The last sub-process of our methodology (Union) match again with the TRIZ Segmentation principle; which consist on facilitating the assembly of the product. It is worth to say that none of the students know the TRIZ methodology and the only possibility to evaluate it is by means of proposing them a methodology inspired on it. Our proposed MDSU methodology is intended to reduce the complexity of making a 3D physical modelling machine and instead manufacturing a 2D physical model and then fold and union it to build a 3D object. We think that this formative manufacturing process is less complex. Furthermore, we are proposing this sequence of applications on TRIZ principles but other sequences might also be explored.

Students were asked to make sketches of the process. They were asked to make at least four alternatives (divergent phase). Figure 3 (right) shows a student work of this experiment. Once they finished the sketches, students were asked to choose on option (convergent phase) and executed accordingly (left figures). They made the experiment with plastic cups commercially available. The tools used by the students were mainly scissors. Finally a questionnaire was applied to the participant students. In the following section we will present the more important results of this experiment.

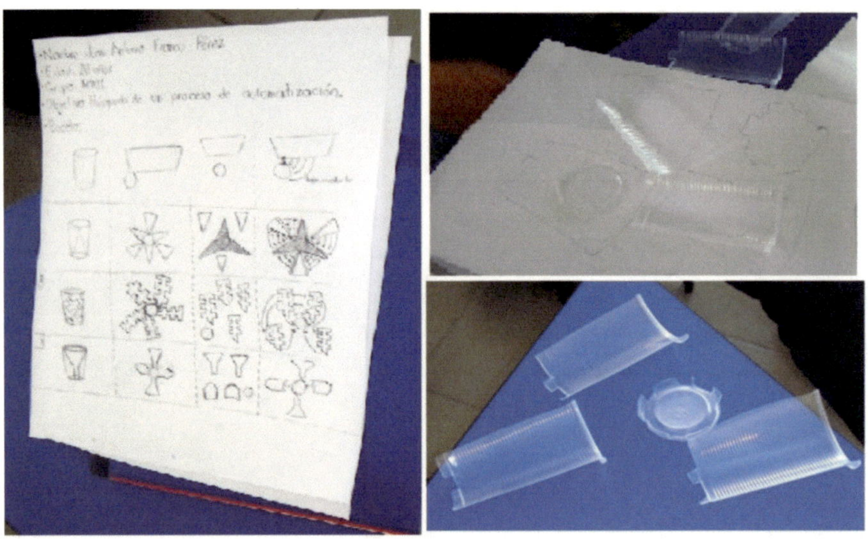

Fig. 3. Pictures from the second experiment.

5.3 Third experiment

The third experiment has the objective to find the best design to the machine automating the MDSU methodology. This experiment was entitled "Automated machine for rapid 3D

physical modelling based on wire bending". Previous experiments were executed with the objective that students have well knowledge of the methodology and its complexity. This experiment was divided in two parts. The first parts consisted in a questionnaire that asked the students about the possibility to automate the MDSU methodology. The technical questions were oriented to design requirements. The second part of this experiment consisted on asking the students a design proposal for the machine. The proposals were done by teams of students. After a self-selected team formation the number of novice teams was 12 and the number of advanced teams was 15. Novice teams were composed by 4 or 5 members. Advanced teams were composed by 3 members. Teams were asked to propose two alternatives, choose and modelling one. Teams had one month to elaborate their proposal in a CAD platform. The experiment was concluded with the presentation of the proposals. Figure 4 shows two proposals from the novice teams and Figure 5 shows two proposals from the advanced teams.
All experiments were executed during the semester August-December 2010. Post-analysis was executed during the first months of 2011.

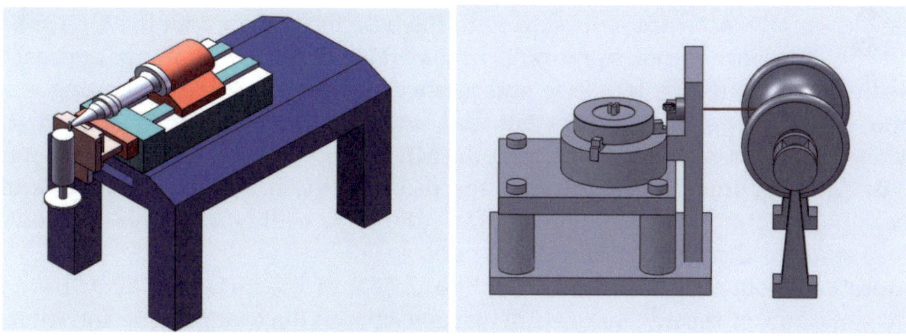

Fig. 4. Examples of proposals from novice teams.

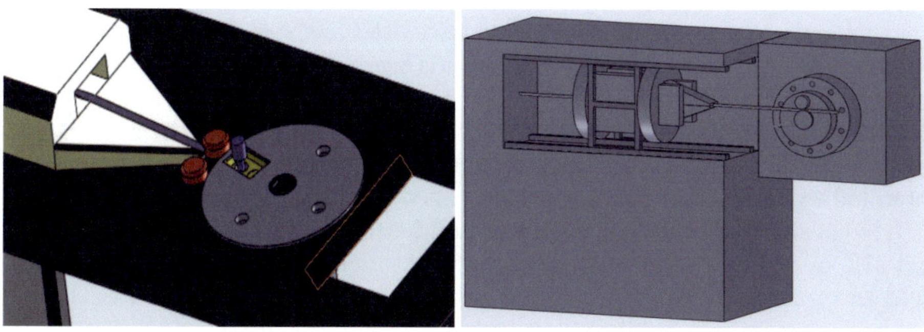

Fig. 5. Examples of proposals from advanced teams.

5.4 The proposed design methodology for rapid 3D wireframing techniques

As explained in Sections 4 and 5, the proposed process to design mechatronic systems for such new rapid three-dimensional physical modelling techniques is the following: first made freeform experiments by hand, second use the proposed MDSU methodology also by hand (the objective of this methodology is to reduce the complexity of the formative manufacturing process for the new rapid three-dimensional physical modelling technique), finally the designer is able to propose a machine (design requirements) for new rapid 3D wireframing techniques.

6. Analysis

6.1 Freeform construction of 3D objects by means of wire bending

The first question was about the thinking process followed by students to manufacture the cup by wire bending. The answer options were: top-base-walls, base-top-walls, base-wall-top, meshed walls-top-base, meshed walls-base-top, 2D meshed, cut sections and union and other. 33% and 30% of the novice and advanced students respectively chosen the top-base-walls thinking process. 28% and 35% of the novice and advanced students respectively chosen the base-x-y thinking process, where "x" and "y" are base or walls. 30% and 26% of novice and advanced students respectively chosen the meshed walls-w-z thinking process, where "w" and "z" are base or top. In general, the common way to think about manufacturing a cup with wire is to build in the following sequence top-base-walls. There are no significant difference between novice and advanced students. If we analyze the statistics deeper, advanced students start with the base more than with the top (35% versus 33%) but the difference is not significant. In general terms three thinking approaches are found with similar statistics: top, base and meshed walls. Something interesting is that 2D mesh and union as well as sections cut and unions were the less considered thinking approaches. This is interested to us because the MDSU approach is more based on mesh and unions. By investigating deeper, this was the result of the instruction that students must build the cup in a continuous way without any cut. But according to our observations a little quantity of students did the job without any cuts.

To the question about the most used basic figure. 36% of the novice students have chosen the circle while 26% of the advanced students have chosen the triangle. Our interpretation is that novice students were inspired by the form of the cup while advanced students were influenced by the knowledge provided in the finite element course they are taking. Another important percentage was assigned to the mix use of basic forms.

Questions related to the criteria that students have used to chose the proposal to be build by hand were applied (divergent phase). The criteria were: design, structural strength, easy to build and material optimization. 30% and 23% of novice and advanced students respectively considered in 100% the design criteria to select their proposals. With respect to the structural support criteria, 36% and 28% of novice and advanced students respectively took into account in 100% this to select their proposals. 17% and 23% of novice and advanced students respectively considered in 100% the easy to build criteria to select their proposals. Finally, 16% and 21% of novice and advanced students consider in 100% the material optimization criteria to select their proposals. From these results, we conclude that structural support was the most important criteria to select their proposals and therefore the most important design requirement for both groups. The novice students considered this criteria more important that advanced students.

To the question about the time spent to manufacture the proposal, novice students spent between 30 to 75 minutes to manufacture their proposal while advanced students spent between 15 to 45 minutes. The results are as expected, advanced students have more manual skills than novice students. It was observed that some students developed a support to manufacture their proposals. Because of that we asked the students if they consider that is necessary some kind of support to manufacture their proposals. 86% and 70% of novice and advanced students respectively think they do not need a kind of support to manufacture their proposals. Our interpretation is that advanced students have more skills and know more tools to develop their proposal and therefore they think in less percentage that they do not need a support.

Because in the first experiment the students manufacture their proposals by hand, a question related to the use of special tools was applied. The results shown that 11% of the novice students and 100% of the advanced students think they need a tool to manufacture their proposals. Furthermore, students were asked about the type of manual tool they need. The results shown that 91% and 94% of the novice and advanced students consider they need tweezers. Other manual tools selected by the students where molds, folding machine, scissors, imagination and none.

An important question about cognition was applied to the students. The question was about generation of ideas about how to improve the formative manufacturing process. 40% and 35% of novice and advanced students respectively feel at 100% that they generate ideas during the experiment. These results show to us that hand manufacturing help to generate ideas (increase creativity).

Recognizing that the final model the students made was a wire-frame model of a 3D object. Students were asked about post-processes needed to finish a 3D object (mock-up). Among the post-processes students proposed are: finishing, structural support, covers, soldering unions, painting, etc. Of the most important for both students were finishing, fixing the support and covering the walls.

One final question about the complexity of the experiment was applied to the students. 8% and 9% of the novice and advanced students considered at 100% the experiment complex. Almost 80% of their appreciation falls between 50% and 75% of level of complexity with a more percentage in the 75%. We conclude that in general students of both levels considered the experiment with some complexity.

6.2 MDSU-based construction of 3D objects by means of wire bending

A set of questions were applied after the students executed the second experiment. There were almost the same questions that in the first experiment. These set of questions will help to us detect the impact of our methodology which is inspired in TRIZ principles. In this section first we will present the questions that are unique to this experiment then in the following section we will present the comparison between the first and second experiments.

The MDSU methodology was executed twice. With respect to the question about the unfolding the options for both executions are: separate the base and sectioned walls, keep the base with sectioned walls, base and walls sectioned, keep some section of the based together to the walls and other. In the first intend, 42% of the novice students chosen the second option while 42% of the advanced students chosen the first option. In the second intend both types of students exchange their choices. In general the first two options were chosen by both types of students. We think that students did not explore other potential possibilities because time restrictions on the experiments. We also asked to the students if the unfolding stage was confused and why. The results showed to us that almost half of the students consider confuse the unfolding stage. The most important reason why they consider that the unfolding is confused is that was difficult to imagine the unfolding.

Another question was related to the complexity of the union sub-process in the MDSU methodology. 29% and 42% of novice and advanced students respectively considered the union sub-process as complex.
In a later section we will compare the first experiment with the second experiment in this parameter. We will also present a comparison of some criteria students took into consideration to select their proposals. Here we present some of the criteria that are not comparable and belong only to the MDSU methodology. Students were asked if once the object was unfolded sectioning was easy. 32% of both type of students considered at 100% that the model was easy to section. Students were asked about other criteria considered to decide their proposals. But they are less relevant for the objective of this study and therefore we will not present them here. Students were also asked about difficulties on the MDSU and related processes but the results will be presented in the following section.

6.3 Comparison between the first and the second experiment

The first compared question is about the geometrical form used in the meshing sub-process (finite element). According to our statistics, most of the geometrical forms used were mix. Both types of students mix more the geometrical figures during the second experiment. In novice students increased 20% while in the advanced students the increase was about 32%. Therefore, there is some evidence that the MDSU methodology help to increase the variability of geometrical forms used in the meshing sub-process. The use of mix forms was superior in the advanced students that in the novice students. The basic form with significant more use after the mix was the circle. This form was reduced about 24% in novice students and 15% in advanced students from the first to the second experiment. The use of triangle was not significantly changed as well as the square. Another important change was perceived in the spiral form; it was reduced about 7% in novice and advanced students. Other forms used where polygon. As we can appreciate the MDSU methodology impact positively the variation of forms more in the advanced students that in the novice. This may be due to the freedom feeling experienced by the advanced students. The triangle is the most suggested form used in finite element theory and according to our statistics it was not changed significantly. This is a positive result because the MDSU does not affect the percentage of use in this form.
When comparing the design criteria for choosing the proposals. The MDSU methodology impacts positively advanced students. For such group of students and at 100%, the criteria is increased by 8% while for novice students the increase is about 7% On the other side, when comparing the easy to build criteria a significant increase is perceived in novice students from 50% to 75%, the increase at 75% is about 29%. At 100%, both groups of students showed an increase when using the MDSU methodology. 15% in novice students and 9% in advanced students. With respect to material optimization criteria, the behavior is of the same kind but at different percentages. The increase at 100% was about 6% in novice students while 8% in advanced students.
The experimentation time had a positive impact also. The more significant increase was shown between 0 and 15 minutes for novice students. By using the MDSU methodology they increase 14% their experiment time. The timeframe between 15 and 30 minutes have had also a positive impact by decreasing less than 10% percentage. Timeframes above 30 minutes had a reduction. These results shown that the MDSU methodology reduces the timeframe spent in the experiment.

Two final important questions were compared. The first was about the complexity of the experiment. At 100%, both group of students considered that the MDSU methodology is less complex than the freeform methodology. Advanced students feel that MDSU methodology reduces the complexity more than the novice students do. The other important question is more related to cognitive processes, specially the creativity. At 100% and 75% (more at 75%) feel that the MDSU methodology reduces the generation of ideas to improve the process.

6.4 Automated machine for rapid 3D physical modeling based on wire bending

The first question of this experiment was about the credibility of both groups of students about the feasibility of automating the MDSU process. 96% and 97% of the novice and advanced students respectively thought that the MDSU process can be automated. Students that do not believe in the automation of the MDSU process justified their answer by stating that such process is not really necessary. We believe that such answers were because they are young and do not foresee potentialities in automating such methodology. From the question that the MDSU process will need a software, 82% and 97% of the novice and advanced students respectively believe that the MDSU process will need a software. Our interpretation of the difference in results is that advanced students have used more software than novice students and therefore they do not see the MDSU process without software. On the contrary novice students have not used so much software as advanced students.

The next series of questions were about the criteria to be considered in the design of a machine that automate the MDSU process. The list of criteria is: design, functionality, manufacturability, execution times, precision, size, feasibility, sustainability, security, easy machine user interaction. For the novice students, the functionality criteria was the most important among all the criteria, followed by manufacturability, execution time, security, then by the design and machine user interaction. For the advanced students the security criteria was the most important followed by the machine user interaction and then by the functionality criteria. Another series of questions were about the criteria students will consider to select if the machine already existed. Novice students selected the precision as the most important factor followed by material optimization and mesh structure. Advanced students selected the precision as well as the most important factor followed by material optimization and mesh structure. The size of the machine was the less important factor.

Another important question was about the students' opinion if the machine will be dedicated to prototypes or final products. 90% of both types of students believe that the machine will serve for prototypes. The last series of questions we asked to the students were about their appreciation of the MDSU-based machine on: reducing complexity, reducing creativity, reducing manufacturing time, and improves work conditions. 49% and 47% of novice and advanced students respectively believe that the MDSU-based machine reduce the complexity of manufacturing 3D Wireframing objects. 49% and 24% of novice and advanced students respectively believe at 100% that the MDSU-based machine will reduce the creativity. This criteria has a more uniform distribution from the 25% to 100%. 75% and 66% of novice and advanced students respectively believe at 100% that the MDSU-based machine will reduce the manufacturing time. Among the set of criteria evaluated in this last series of questions this is with the more believability. The last criterion in this series is about the improvement of work conditions. 60% and 66% of novice and advanced students respectively believe at 100% that the MDSU-based machine will improve the work conditions.

The last series of analysis executed in this experiment was about the students' proposals. In section 5.3 we have provided some examples of MDSU-based proposals. Students were asked to make their proposals without design restrictions; except that all must met the MDSU methodology. Novice student proposals were more oriented to the mesh sub-process while advanced student proposals met well the MDSU process. A comparison among the proposals was executed taking into account the following sub-processes: meshing, folding, sectioning, cut, union, wire size, machine size, continuous feeding, straighten system, diversity of forms capability, folding ranges. It was noticed that cut and wire size was met by all the proposals in both types of students. 92% and 100% of novice and advanced student proposals respectively are machines of considerable size. 83% and 93% of novice and advanced student proposals respectively consider continuous wire feeding. 58% and 53% of novice and advanced student proposals respectively met with the meshing requirement. 67% and 87% of novice and advanced student proposals respectively me the folding requirement. 50% and 73% of novice and advanced student proposals respectively met the sectioning requirement. In general advanced student proposals met the MDSU requirements better than the novice student proposals; except the meshing requirement but not for more than 5% of difference.

7. A first prototype

The current prototype was conceived in a multidisciplinary way almost following the concurrent engineering approach. Three different specialties were participating: industrial design, mechanical engineering and electronic engineering. A professor from each specialty and a Master of Science student from each discipline participated. We had meetings every three weeks. During the meetings the final mechanical design was decided. Once the mechanical design was decided, the electronic design starts to automate the machine. Our first prototype was developed during these meetings. Figure 6a shows the first prototype we developed. There were no design process followed for the first prototype. After the first proposal we carried out some simulations in Rhino™ to detect possible problems. We found some problems that were corrected in the second prototype shown in Figure 6b. The prototype shown in Figure 6a can only make wire bends from ±90°. Figure 7 shows a picture of the real prototype as shown in Figure 6b. Some first tests were executed with basic 2D figures. Figure 8 shows some basic geometrical forms done with our first prototype. The

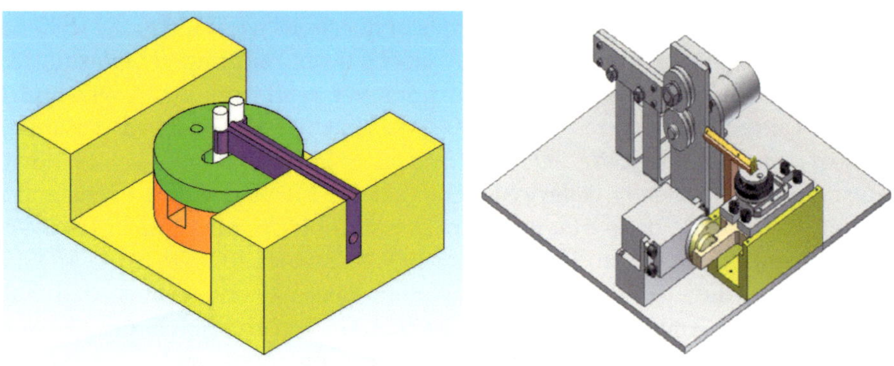

(a) (b)

Fig. 6. First prototypes.

reader can appreciate that many changes must to be done before our first MDSU-based prototype machine will be achieve. With the design process we propose here and its results we will propose a new machine that meets the MDSU requirements. Due to intellectual property rights we cannot show more details of our first prototypes.

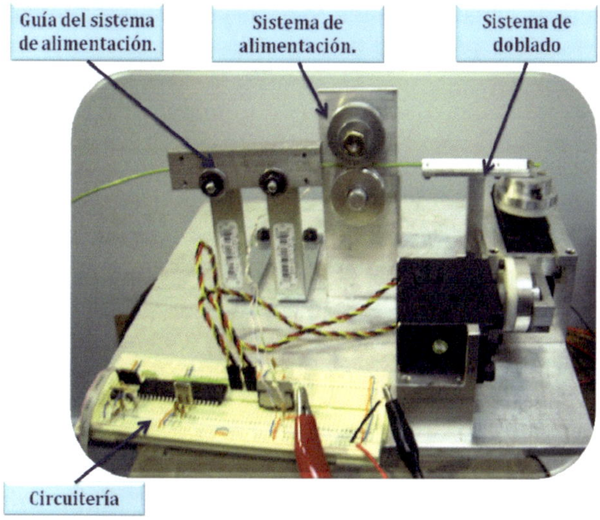

Fig. 7. Picture of the first prototype.

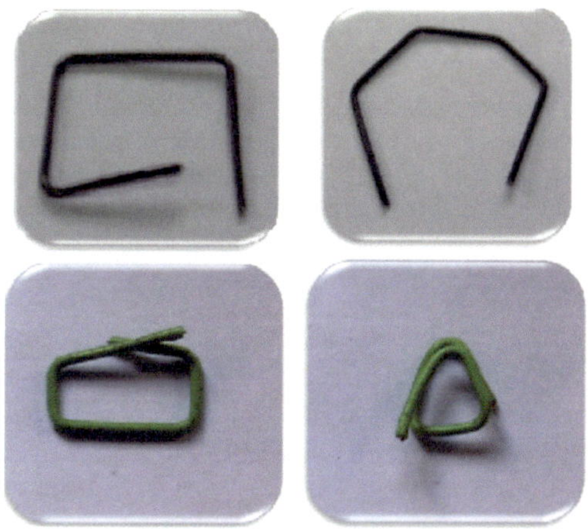

Fig. 8. Basic geometrical forms manufactured with our first prototype.

8. Conclusions and future work

In this chapter we have proposed a methodology inspired in TRIZ principles to design mechatronic systems for a new rapid 3D physical modeling technique based on formative manufacturing processes. This chapter has several contributions that will be outlined in the following. First, it presents a match between the most important engineering design frameworks: engineering design process, product lifecycle management and project management. It also adds an experienced course published previously (Cárdenas 2009), the INNOWIZ framework and the TRIZ framework. Second, focused in the first two stage of such engineering design thinking an innovation funnel was proposed where the application of rapid prototyping techniques is outlined. Traditionally, rapid prototyping techniques are used once the concept is well defined (after detailed design) but previous research has proven to be used in early stages of design (e.g. conceptual design phase). Third, formative manufacturing processes are proposed as a new paradigm to explore the design of new rapid 3D physical modeling techniques. Fourth, a new rapid 3D physical modelling technique based on a particular case of formative manufacturing processes such as wire bending and inspired in TRIZ principles was proposed. The methodology is called MDSU (MUSU in English). Fifth, a design process for mechatronic systems of rapid 3D physical modeling techniques was proposed and consists of three stages. Making the 3D object by freeform hand, and then making the 3D object following the MDSU methodology, finally proposing new machines that automate the MDSU methodology. This new paradigm promise potential benefits in favor of sustainability issues. This design methodology does not follow the traditional engineering design path but presents an extension in the problem definition stage. Because the complexity of automating handwork operations the problem (stage one in the engineering design process) must be defined by experiencing the two first stages of the proposed design process. In other words, the first two stages of the proposed design methodology must be used in the problem definition stage of the engineering design process. The third stage consists of the rest of the engineering design process from researching the problem. Now, according to our experiments, the MDSU methodology increases the use a mix of geometrical forms in both novice and advanced designers. In general, the design criteria in both types of designers are increased if the MDSU methodology is used. Another positive impact of the MDSU methodology was the execution time. Execution time was shown to improve. It is expected that machines using this methodology will perform better in terms of time. The MDSU design methodology reduces the complexity to manufacture rapid 3D physical models, specifically rapid 3D wireframing objects. We have some evidence that, MDSU-based machines will reduce the time and complexity to manufacture 3D wireframing models but also will reduce the capacity to generate ideas. On the contrary, it will promise to improve work conditions. Finally, advanced student proposals met better the MDSU design requirements that novice student proposals except for the meshing sub-process. Finally, we provide advancements of our first prototype machine which by now process only basic two-dimensional figures.

As a future work, we are planning to finish the first prototype that follows the MDSU methodology. It will surely comprise hardware and software advancements with respect to the Figures 6b and 7 shown previously. We also plan to continue the evaluation of the same parameters used to conclude in this chapter and a deeper analysis of the current results. Finally, formative manufacturing processes are a wide potential area that has been less exploited and the different types of material might open new potential possibilities. We will explore the application of our MDSU process to new rapid 3D physical modeling techniques not only based on one-dimensional materials but two- and three-dimensional.

9. Acknowledgments

Authors would like to thank the support of CONACYT, the Mechatronics department at Tecnológico de Monterrey - Campus Querétaro, and the Master in Manufacturing Systems at the same Institution. This work has been supported by the Distributed and Adaptive Systems Lab for Learning Technologies, DASL4LTD (C-QRO-17/07) and by the Innovation in Design and Manufacturing Research chair, both from Tecnológico de Monterrey - Campus Querétaro.

10. References

Bruno, F., Giampŕ, F., Muzzupappa, M., Rizzuti, S. (2003). A Methodology to Support Designer Creativity During the Conceptual Design Phase of Industrial Products. Proceedings of the 14th International Conference on Engineering Design ICED03.

Buchal, R.O. (2002). Scketching and Computer-Aided Conceptual Design. The 7th International Conference on Computer Supported Cooperative Work in Design, 2002.

Buswell, R.A. (2007). Applying future industrialized processes to construction. IN: Walker. N. (ed.). Proceedings of CIB World Building Congress 'Construction for Development' 14-17 May 2007, Cape Town, South Africa.

Buxton, B.(2007). Sketching user experiences - Getting the Design Right and the Right Design. Springer (1st Edition).

Cárdenas, C. (2009). Social Design in Multidisciplinary Engineering Design Courses. 39th IEEE/ASEE Frontiers in Education Conference, San Antonio Texas, USA, October 18-21.

Cárdenas, C., (2011). A Multidisciplinary Approach to Teach the Design of Socially Relevant Computing Systems for Social Change. International Journal of Engineerinf Education, Vol 27. Issue 1, pp. 3-13.

Cha, J., Jardim-Gonclaves, R., & Steiger-Garcao, A. (2003). Concurrent Engineering. Taylor & Francis (1st Edition).

Chua, C. K., Leong, K. F. & Lim, C. S. (2010). Rapid Prototyping: Principles and Applications. World Scientific Publishing Company. (3 edition).

Crawley, E., Malmqvist, J., Ostlund, S., Brodeur, D. (2010). Rethinking Engineering Education: The CDIO Approach. Springer (1st Edition).

Demaine, E.D. & O'Rourke, J., (2007). Geometric Folding Algorithms: Linkages, Origami, Polyhedra. Cambridge University Press, July 2007. xii+472 pages. ISBN 978-0-521-85757-4.

Dym, C.L., Agogino, A.M., Eris, O., Frey, D.D., Leifer, L.J., (2006). Engineering Design Thinking, Teaching, and Learning. IEEE Engineering Management Review. Vol. 34, Issue 1.

Gibson, I., Rosen, D.W. & Stucker, B. (2009). Additive Manufacturing Technologies: Rapid Prototyping to Direct Digital Manufacturing. Springer (1st Edition).

Hopkinson, N., Hague, R., & Dickens, P. (2006). Rapid Manufacturing: An Industrial Revolution for the Digital Age. Wiley. (January 16, 2006).

Jenkins, D.L. & Martin, R.R. (1993). Importance of free-hand sketching in conceptual design: Automatic sketch input. ASME DES ENG DIV PUBL DE., ASME, NEW YORK, NY (USA), 1993, vol. 53, pp. 115-128.

Kamrani, A. K., & Nasr, E.A. (2010). Computer based design and manufacturing. Springer (1st Ed.).

Kamrani, A. K., & Nasr, E.A. (2010). Engineering Design and Rapid Prototyping. Springer (1st Ed).

Kiritsis, D. (2011). Closed-loop PLM for intelligent products in the era of the Internet of things. Computer Aided Design. Vol. 43, Issue 5 May 2011, pp 479-501.

Kupka, L. (2010).Physical modelling aids feeder design and assessment. Glass International. Vol. 33, (2010), pp. 31-35.

Lessard, C., (2007). Project Management for Engineering Design. Morgan and Claypool Publishers (1st Edition).

Orloff, M.A. (2010). Inventive Thinking through TRIZ: A Practical Guide. Springer (2nd Ed.).

Pahl, G., Beitz, W., Feldhusen, J., & Grote, K.H. (2007). Engineering Design: A Systematic Approach. Springer (3rd. Edition).

Plattner, H., Meinel, C., & Leifer, L. (2010). Design Thinking: Understand - Improve - Apply. Springer (1st Edition).

Saaksvuori, A., & Immonen, A. (2010). Product Lifecycle Management. Springer (3rd Ed.).

Schrage M. (1993). The culture(s) of prototyping, DesignManagement Journal, 1993, 4 (1) : 55-65. In Broek Johan J., Sleijffers, Wouter, Horvath, Imre, 'Using Physical Models in Design' Dept. of Design engineering, Delft Univ. of Technology.

Simondetti, A. (1998). Rapid Prototyping Based Design: Creation of a Prototype Environment to Explore Three Dimensional Conceptual Design. Cyber-Real Design Bialystock (Poland), 23-25 April 1998, pp. 189-203.

Smith, P.G. & Reinertsen, D.G. (1998). Developing Products in Half the Time, John Wiley and Sons (2nd Edition), New York, 1998.

Suh, S.H., Kang, S.K., Chung, D.H. & Stroud, I. (2010). Theory and Design of CNC Systems. Springer (1st Edition).

Zailani, S., Rajagopal, P., Jauhar, J. & Wahid, N.A. (2007). New product development benchmarking to enhance operation competitiveness. International Journal of Services and Operations Management. Vol. 3, no. 1, (2007), pp. 23-40.